The Sufficient Community

S0-BAP-192

CHRIS WRIGHT

The Sufficient Community

Putting People First

GREEN BOOKS

Published in 1997 by
Green Books Ltd
Foxhole, Dartington,
Totnes, Devon TQ9 6EB

Copyright © 1997 by Chris Wright

The right of Chris Wright to be
identified as the author of this work
has been asserted by him in accordance with
the Copyright, Designs and Patents Act of 1988

All rights reserved

Book design and production by Janet Allan
Cover illustration by Clare Allan

Typeset in 11 on 13 Bembo
by Koinonia, Bury

Printed on acid-free paper
by Biddles Ltd
Guildford, Surrey

A catalogue record for this book
is available from The British Library

ISBN 1 870098 68 4

Contents

Acknowledgements

The author would like to thank the following who have kindly given permission for the use of quotations from copyright material: Oxford University Press for James Lovelock's *The Ages of Gaia* 1990; Picador for Jung's *Man and his Symbols*; Peter Fraser and Dunlop Group Ltd for Arthur Koestler's *The Ghost in the Machine*; BBC Worldwide for Michael Andrews' *The Birth of Europe*; A. P. Watt Ltd and Dr Patrick Nuttgens CBE for *The Home Front*; Rider and M. Scott Peck for *The Different Drum*.

Introduction

And so each new venture is a new beginning.
A raid on the inarticulate with shabby equipment is
always deteriorating.   T. S. ELIOT

A few kilometres north of the Angouleme to Limoges high way lies Oradour-sur-Glane. On 10 June 1944 – four days after allied troops set foot in Normandy – the village was surrounded by men from the 3rd Company of the 'Fuhrer' Regiment of the SS. The inhabitants were herded together, divided into six groups and gunned down. Six hundred and forty-two died. The houses and church, where the corpses lay, were set ablaze. Only one woman and five men survived.

Why the massacre happened remains unclear but, when he saw the ruins, General de Gaulle decreed that they should be preserved untouched as a visible testament to what France had suffered under the Occupation. A new village was built alongside and the site of the old has become a shrine visited by hundreds of thousands each year.

Only the shell of a once thriving community remains. Small plaques have been attached to the broken walls of each house giving the family name and their occupation: homely trades such as dress-maker, mechanic, shop keeper, labourer... Up the main street the tracks of a tramway still run, complete with overhead cable and antiquated ceramic insulators. They, more than anything, fix the place in time and make it so much more than an abandoned village. For this place is truly empty. It has been literally emptied of life.

Coming away from Oradour I felt only an overwhelming

pointlessness. There seems to be no room for outrage or anger – the enormity of what happened precludes that – just a wondering at this strange creature that oscillates between joy and desolation, that can find within itself the vision to create great works and just as easily to bring them crashing down. In Oradour we can see the two sides of our nature imprinted in the ground. On the one hand, the creativity that is everywhere apparent in ordinary life; on the other, the life-threatening power of the impersonal forces that we seem so ready to surrender ourselves to.

That is the abiding truth that the empty streets of Oradour bear witness to, and it is a truth that we forget at our peril. At every point in history the balance is delicate, and no more so than today, when events appear to be running beyond the control of any individual. Whether it be someone fighting to avoid the repossession of their home or coming to terms with redundancy, a group protesting at proposed motorway plans or factory farming, or the threats of global warming and the thinning of the ozone layer, we, as individuals, are increasingly powerless in the face of institutional forces we do not understand, let alone control.

Conventional wisdom suggests that the only remedy is more of the same. Progress comes through economic growth and we must therefore become ever more competitive. The nation – or group of nations – that develops the most effective technology, invests in it and then creates the markets for what it produces, will assure for its peoples the greatest returns on their collective endeavours in terms of material standards of living, healthcare, educational opportunity, leisure and recreational facilities, etc.

There is the brave new world of information technology, the prospect of cheaper and faster transport and communication networks, the potential in robotology for overcoming skill shortages and protecting workers from hazardous environments, the promise of genetically engineered breeds of high-yield, pest-resistant crops able to flourish in the most inhospitable environments. And, forever beckoning, lies Space – the last frontier – with its expectations of untold riches. It is there for the taking if only we have the courage to reach out...

This relentless pursuit of new markets is being conducted by multinational companies, many of which have annual turnov┊ excess of nation states as rich as Switzerland. Increasingly, tran┊ systems are national – if not intercontinental – in orientation, rather than a means of getting from one locality to the next. The emphasis is on a flow of goods and services rather than a flow of people. Factories close or get moved elsewhere in response to wider political or economic considerations, throwing thousands out of work. People have to move more and more frequently in pursuit of promotion, security, a new challenge. Amenities such as shops and cinemas become ever bigger and are sited in places that make car ownership essential. Time itself becomes telescoped in the increasingly frenetic race to keep ahead. People become less and less important. In fact, they become a nuisance, infecting the lavish models that seek to squeeze the last drop of economic efficiency from plans to further enhance competitiveness.

What is also seldom mentioned – because it is unthinkable – is the possibility that if the economy fails to deliver the goods, or that we become one of the also-rans in the economic Olympics. Significant poverty already exists in our society. Oxfam, the Third World charity, now provides aid to the poor in its own back yard, recognising that poverty is a direct contributor to major social evils such as physical and mental ill health, homelessness, marital disharmony and family break-up. In that sense the economy is already failing and the situation is getting worse.

While statistics appear to show that, for those of us still in the race, our standard of living continues to rise, we know that the quality of that life is deteriorating. There is a feeling that we live in an ever more violent, less caring society. Petty irritations, from traffic jams and aggressive driving to the expense and effort of getting even the simplest repair jobs done, bedevil us and mock the power that wealth has put in our hands. Beggars on the streets affront us and there is no escape from the crowds.

As society becomes ever more complex, specialised and impersonal in the pursuit of economic growth, so does the energy needed to hold it together. The public's preoccupation with muggings, robbery and vandalism and its apparent desire to barricade itself ever more securely within its own homes, speaks more of life in the Dark Ages than of a sophisticated civilisation enjoying one of its longest periods of continuing peace. Something is clearly wrong and there is little agreement on how to respond to the challenge of individuals and groups who exhibit little commitment to, or ownership of, the society we share. Consumer terrorism, bombings and hijacks all reinforce the sense of fragility that turns life into a game of Russian roulette, with the individual surrounded by a sea of potentially hostile and unknown faces. The only way of keeping the social fabric together in such circumstances is likely to be through increasing repression.

From such images many people, from a variety of political, religious and philosophical viewpoints, are now concluding that the way we currently live simply doesn't work. The language in which the questions are being posed may be very different, but there is an increasing consensus about the need to explore alternative ways of relating to one another if we are not to lose our humanity in the face of such onslaught. What is also common is that the vision of a new social order is expressed through the notion of *Community*. Unfortunately, it is a concept that has become devalued with usage, obscuring as much as it reveals. Indeed, for many it represents a chocolate-box study of a past that never existed. For the word to offer a genuine signpost to the future requires a fundamental re-examination of its many meanings.

The problem is that there is no generally accepted framework within which ideas (ideals) about 'community' can be debated. The early writings that came to underpin what we call capitalism, for example, were   essentially concerned with the moral, social and economic role of the individual within society. People mattered. It is ironic that such thinking provided the bedrock of assumptions that has led directly to our current dehumanised concepts of 'the market place' and 'economic growth', concepts so commonplace and apparently self-evident that it can sometimes take an effort of will to remember that there might be other visions.

Over two centuries have passed since Adam Smith wrote *The Wealth of Nations*. In that time no society has ever been truly capitalist. Those ideas have nevertheless fired generation after generation to find their own truth in what has become a unified thought system. Until the ideals on which community living might be based can be established as coherently and effectively, we are unlikely to experience the upsurge of experimentation and discussion that characterised the early years of the capitalist era and which can still produce passionate debate today. It is that energy that we urgently need to harness. We need to find a way to reclaim our humanity if the impersonal forces of the money economy are not to overwhelm us as effectively as the Gestapo destroyed Oradour.

1

The Pursuit of Happiness

Happiness is the meaning and purpose of life, the
whole aim and end of human existence.

ARISTOTLE

Self-development is big business. Individuals and groups, offer-ing a range of therapies to keep body and mind healthy, are sprouting up everywhere. And with them come regimes and theories about every aspect of human functioning. From psychoanalysis to aerobics, from primal integration to diet plans, and from transactional analysis to allergies, people are feeling the need to grapple with personal issues rather than merely enduring them.

'Hothousing' could be seen to be taking this trend to its logical conclusion. The race is on to produce 'super people', able to get on in society and be super achievers in both work and leisure. Why wait to sort out blockages in adulthood when gifted individuals can be produced by creating the right environment in the early years? It is never too early to start, and some exponents require the co-operation of both parents while the child is still *in utero*. The nation to emerge pre-eminent in the twenty-first century will be the one that succeeds in the hothousing experiment!

On the other hand, if you ask parents themselves what it is they are hoping to provide for their offspring the answer is invariably 'happiness' − and that dream is as old as time itself. Fame and wealth would be nice, it seems, but essentially we want our children to be happy. The American Declaration of Independence begins 'we hold these truths to be self evident, that all [people] are created equal; that they are endowed by their Creator with certain inalien-able rights; that among these are life, liberty and the pursuit of happiness'. What then is 'happiness' that it features so prominently in our hopes and aspirations?

One way into that question is to ask any group of people who they would choose to join them on a desert island and why. A list of criteria could be drawn up without much difficulty and there would probably be a large measure of agreement both within the group and between others doing the same task. What they would really be exploring is what makes life worth living. It's a fun exercise that usually starts off with a light-hearted look at the need for survival – 'whatever else we need, we need someone who can cook!'. The issue of the balance between the sexes soon raises its head and a degree of fantasy frequently enters the discussion. Before long, however, the need for getting on together and for pooling skills and resources predominates. The prospect of spending months, perhaps years, together has a tendency to focus the mind.

The list that emerges forms a 'hierarchy' of needs, as described by Abraham Maslow,[1] that range from the basic, physiological requirements for food and drink, through the desire for security and stability to higher-order and increasingly complex aspirations such as belongingness, self-respect and, ultimately, the realisation of oneself as a unique individual. Fundamental to that shift from the meeting of animal to less tangible needs is a move from the material to the spiritual; an increasing sense of awe and wonder at the universe and one's own, individual existence within it, and a consequent feeling of being at one with creation.

It is a hierarchy because lower-order needs (physiological) must be satisfied before those higher up the ladder. Someone who is starving is unlikely to have much self-esteem. If that situation persists for any length of time then, even when circumstances improve, it may be difficult for the individual to realise higher-level needs. The survivors of the Holocaust, for example, have spent a lifetime trying to come to terms with what happened to them and, even after fifty years, suicide remains the only solution for some. Children, whose development depends on having adults around who are themselves having their higher-level needs met, can be hit particularly hard by periods of deprivation of any sort. It is salutary to reflect that the child victims of drought and civil war will be stunted physically and emotionally should they survive into adulthood, with a consequent limiting of their own ability to be parents.

What such a schema highlights is the importance of other people in our lives. Even our most basic 'animal' need for food and drink is met in a social way. We go to extraordinary lengths and expend vast amounts of creative time and energy on assembling inviting and appetising arrangements of basic raw materials. We do it for others to enjoy but, in the process, to have our own higher-level needs of love and esteem met and reinforced. If that effort is taken for granted, however, the opposite effect is achieved. Relationships have to be living, growing interactions if they are to be liberating not limiting. Other people and the quality of the relationships we enjoy with them are thus of vital importance to how we feel about ourselves and the outside world. In a society that puts 'self' centre stage and fosters a belief in rugged individualism and 'looking after number one', that rather obvious statement can be lost in the chorus of appeals to please and pamper ourselves.

Whatever 'happiness' may be, it can never be a constant state. We need to have been 'unhappy' to recognise and appreciate contentment and ecstasy. Friendships are as likely to provide pain as pleasure but it is through our deepest and most intimate relations that we are able to explore, understand and come to terms with who we are. They also provide the opportunities for growth. Whether we capitalise on them is up to us but life is an unfolding process, a journey rather than a resting place, and to stand still for any length of time is as harmful to the psyche as dashing from one excitement to the next.

Almost anything we do has a social context, and other people, directly or indirectly, offer us opportunities to satisfy our needs. At the very least, other people offer us affirmation. In telling someone our joys and our sorrows we make them real. In doing something with them or for them we are utilising and demonstrating our competence. Without the other the self remains untested, potential rather than actual. To realise our 'selves', therefore, we must put ourselves in the hands of others. We have to risk all to gain all.

'Self' versus 'Other'

It is important to emphasise that, while others are crucial t(
sense of a developing identity, it should not be assumed that life
consists of nothing but relationships or, more particularly, our most
intimate relationships. Pursuits and interests are often undertaken
in isolation from others, and having time to oneself is necessary to
absorb and assimilate life's lessons. Nevertheless, what we think, do
or make can ultimately be validated only through the other. Using
our talents, expressing and being ourselves is an exchange with our
environment and, most importantly, the people in it. It is through
such interactions that we learn the limits of what we, as individuals, can be.

Our materially affluent society takes the meeting of basic, physiological needs for granted and is more concerned with what it
perceives to be the higher reaches of the hierarchy. Self-realising
people are, to paraphrase the psychotherapist Carl Rogers, people
who are in the process of becoming: a voyage that requires an
acceptance of, and trust in, one's whole organism. 'Consciousness,
instead of being the watchman over a dangerous and unpredictable
lot of impulses, of which few can be permitted to see the light of
day, becomes the comfortable inhabitant of a richly varied society
of impulses and feelings and thoughts, which prove to be very
satisfactorily self-governing when not fearfully or authoritatively
guarded.'[2]

It is one of the central paradoxes of human existence that the
more we seek to realise our own, individual potential (self-
realisation), the more we need other people. On the face of it, that
is sufficient explanation for the fact that the world isn't full of self-
realised beings, at peace with themselves and their neighbours; and
why the Oradours and the Sarajevos are more common than the
Taizes or the Findhorns. Other people, far from assisting us towards fulfilling our potential, always appear more likely to limit
our freedom of action, our freedom to do as we please. One of the
consequences of the communications revolution, for example, is
that the number of people who can get in touch with us at any
given moment, and who believe they have the right to do so,
continues to mushroom. As a result we spend less and less time in

the present – being with ourselves – and more and more time in the past or the future depending on whom we are momentarily in communication with.

We have plenty of relationships, but few of them are of the kind that actively contribute to our self-realisation. The majority of our daily interactions are on the basis of roles. We relate to other people as 'a teacher' or 'a shop assistant', rather than as real people with an existence as complex as our own. So much of our social behaviour exists on this plane, fleeting and impersonal, that it is easy to forget what true human contact is like. Even our friendships are largely social in the sense that we choose people from within the same income and interest groups, people we can sit down and pass the time with. We seek to be diverted not challenged.

Given our increasingly specialist knowledge and limited circles of experience, that means selection from a narrowing spectrum of the population. We *do* very little with others; co-operation usually comes only after friendship has been well established. The effort of keeping in touch with networks that may span the country, if not the world, can become so immense that we opt out, choosing to sit in front of the box. Within the home itself, where our closest and most meaningful links should be, life is so dominated by the daily demands of work and school that we meet only for limited and predictable periods, at times when we are recovering from the slings and arrows suffered elsewhere; spontaneity has vanished. Conflict is everywhere and the support systems that might help us cope are increasingly fragmented and fragile. It is hardly surprising that there is so much pressure on our marriages, or that they break down as frequently as they do.

Human relationships that are likely to encourage self-realisation must be based on equality and mutuality. Yet, in our society, the emphasis is on rising above the masses, on becoming 'someone'. It is as if the only way we can be sure of our significance is to deny the importance of everyone else. How then can true equality between individuals exist without everyone being lost in the common herd? The answer is that, within the context of the world we currently inhabit, the opportunities for resolving the dilemma are . We have become partial beings and the pursuit of power is a compensation for our lost selves. It is a vicious circle; the more we attempt to exert our individuality by trying to make our self heard above the hubbub of everyday life, the more self-contained and cut off we become. We gradually lose the ability to see 'others' as anything beyond the role they perform in our lives. Without the kind of relationships that lead to self-realisation, we have to shore up our increasingly fragile sense of 'self' by getting even further ahead.

The first step back on the road to sanity is to recognise that we need a network of relationships around us that are open-ended and full of potential meaning; the antithesis, in fact, of a life dominated by roles which, by their nature, are closed and self-limiting. Such networks are very difficult to create and sustain in the large-scale organisational settings that dominate our social structure and where change and mobility are built into the fabric. They are, in fact, possible only in settings that we would normally describe as communities.

A child of the times
A community cannot just be created, however. At the very least, the circumstances in which we find ourselves play a crucial role in determining the kind of relationships that are available. A period of prolonged famine, for example, can lead people to abandon any form of sociability, including raising the next generation. Colin Turnbull's study of a tribe of nomadic hunters[3] shows what the bottom line is like and how far removed it is from the values we would normally ascribe to humanity. Driven out of their traditional hunting grounds in Central Africa, the Ik had been forced into the barren mountains of Northern Uganda and were barely able to sustain themselves. Overnight, they lost all the yardsticks by which they usually measured themselves. Hunting requires very different skills from farming, and the ages-old antipathy between the two lifestyles suggests that attributes considered important in one are likely to be thought irrelevant in the other. Even if they had made the transition successfully, the land was so poor that life would always be precarious, leaving little room for any sense of achievement or progress.

Perhaps not surprisingly, the Ik gave up. Little cultivation took place and people lived off the land. Life became a constant round of stealing from even one's closest relatives while, at the same time, concealing from them what little one had managed to gather or grub up oneself. Elaborate subterfuges were practised to protect the location of a root or seed not yet ready for eating. Meanwhile, children and the elderly were left to fend for themselves. Individual survival was the sole criterion by which to live and judge relationships. As a people the soul had gone out of them and nowhere was that more apparent than in the quality of their relationships.

By contrast, the following description of lifeboatmen at the turn of the century reveals a different set of priorities:

a fearful snowstorm, blowing across the Channel, raged on the flat, sandy coast of a tiny village in Kent, and a small smack, laden with oranges, stranded on the sands nearby. In these shallow waters only a flat-bottomed lifeboat of a simplified type can be kept, and to launch it during such a storm was to face an almost certain disaster. And yet the men went out, fought for hours against the wind, and the boat capsized twice. One man was drowned, the others were cast ashore. One of these last, a refined coastguard, was found next morning, badly bruised and half frozen in the snow. I asked him, how they came to make that desperate attempt. 'I don't know myself,' was his reply, '*There* was the wreck; all the people from the village stood on the beach and all said it would be foolish to go out; we never should work through the surf. We saw five or six men clinging to the mast, making desperate signals. We felt that something must be done, but what could we do? One hour passed, two hours, and we all stood there. We all felt most uncomfortable. Then all of a sudden, through the storm, it seemed to us as if we heard their cries – they had a boy with them. We could not stand that any longer. All at once we said, 'We must go!' The women said so too; they would have treated us as cowards if we had not gone, although next day they said we had been fools to go. As one man we rushed to the boat, and went."[4]

In terms of material well-being there was probably not much to choose between these two peoples: both lived marginal existences. What separates them is a set of mutually supportive relationships that allowed individual needs to be met. 'We must go!' It is a desperate cry but one that speaks volumes about the integrity of the individuals involved. When any community faces a crisis it depends on the moral strength that is contained within. Failure to meet a crisis is corrosive. The structure may look strong from the outside, but ultimately it may collapse of its own accord or at the slightest external pressure.

The level of common interest and fellow feeling demonstrated by the lifeboatmen of Kent a century ago is something that can appear entirely absent from our cities today: a child repeatedly abused in the next house, a woman mugged in front of a crowd of commuters. We have increasingly withdrawn into ourselves as the identification one with another – the sense of belonging to a community – that might have prevented such outrages has withered away. How far our society has progressed along the road towards the self-centred deadness displayed by the Ik is a matter of judgement, but what is clear is that no culture can survive once the basic instinct for co-operation has been squeezed to the margins. Only coercion remains to force individuals to act together for the sake of production, the economy or the greater good of the state.

How we relate to one another is thus a good indicator of where our priorities lie. The increasing emphasis on roles and interacting with people we hardly know suggests that the balance is tilting towards co-operation that is externally ordered and away from that which springs from a spontaneous, internalised wish to act together. The implications for individual self-realisation, let alone a society's ability to surmount crises, should be obvious.

Can the slide be halted? The simple answer is that yes, of course it can. How we relate to each other is a matter of choice – *our* choice. If we want to change the way we relate to one another we can. But we need to recognise what we are up against. That means accepting some of the limitations imposed by the fact that we are born into a world that is already determined in many key ways. Who we may be capable of becoming – and what we can therefore bring to relationships – depends on a complex interweaving of forces that guarantees our unique potentialities while also fixing our room for manoeuvre.

To the limited extent that we understand these matters, that balance initially hinges on the nature of the human psyche (its drives, motivations – both conscious and unconscious – and, ultimately, its purpose) and the physical and chemical universe it occupies. That much we share in common. Thus, our world is characterised by the fact that we are creatures who attain, on average, a height of between 1.6 and 2.2 metres, stand upright on two legs, have physiological requirements that include eating, sleeping, defecating, etc. and, perhaps most importantly, are born dependent, grow to full maturity and die within a time-scale of three score years and ten. Even our sense of symmetry, which tends to be vertical, comes from the symmetry of our own bodies (two eyes, midline nose, etc.) and the effect of gravity on nature in general. Our physical existence is also responsible for our sense of scale and the restrictions we have to impose on ourselves and our environment to survive in terms of, for example, acceleration, heat and weight. The rhythms we take so much for granted – day and night, the lunar month and solar year – have subtle influences on us, as well as providing a context for our existence. That these fundamental fixed points of existence are ignored is evident in all aspects of our daily life, from the vast, impersonal housing estates that are being thrown up everywhere to the excesses of the motor car culture. They could only have come about because people have lost their sense of what it is to be human.

We also have to accept the limitations of our own personal inheritance, the hand that has been dealt to us through the genes that determine our appearance, gender and aptitudes, as well as the parents who gave them to us and who influence, through their own personal example, much of how we play our cards in practice. Important as these factors are, however, they are more than balanced by the role played by those elements we humans impose on ourselves. Growing up is like walking on to a stage set, being presented with a script and told to get on with it. We learn our 'part' and interact with a cast limited in both number and scope for action.

From the moment we are born we absorb information through all our senses. Observing how people relate to one another, who is considered important and how they achieve that status, shows us the subtle nuances of language, what it is appropriate to say and how to say it. We also take in the conventions of dress and gesture. And we do it in a largely uncritical manner because that is the way it is; self-evident, like accepting that a washing machine is for cleaning clothes rather than cooking the dinner. This all-embracing social and cultural milieu acts in three interrelated but quite distinct ways:[5]

1　how individuals relate to one another through the sharing of a common language, mannerisms and dress – the plane of social behaviour
2　how individuals are encouraged to conform to accepted behaviours by the 'structural' framework that exists in terms of laws and the range of social organisations that ultimately derive their authority from the law – the plane of social structure – and
3　what individuals believe to be important – the plane of values.

It is usually the tension that exists between these three levels that provides the dynamic for change.

All the World is a Stage – the plane of social behaviour
Social (as opposed to personal) behaviour influences our ability to relate to one another on a day-to-day basis. As a tourist, unable to speak the language, one quickly realises one's limitations but, even *within* a culture, obstacles are everywhere. A feudal noble and a peasant would find it difficult to develop a mutually satisfying relationship because of the assumptions each has about the other, attitudes that are reinforced by language, dress and mannerism as much as by inequalities in wealth and power. These assumptions allow us to categorise people – and all too often to dismiss them – rather than get to know them. As such they pose real barriers to communication and the results are only too evident in the annals of history. It is easier to see such disparities from a historical perspective but they exist today both within our society and between cultures. They are like time bombs ticking away.

That is not to argue that we should all talk, act and dress the same, because it is only through the challenge of diversity that we can review our assumptions and modify them in the light of experience. Uniformity and conformity produce a mausoleum of the living dead, people who are quite unable to respond to the challenges of life because life itself has become a series of stereotypical responses endlessly repeated. Difference, paradox and conflict lead to a redefining of experience and the emergence of new meaning. The great hope of humanity is the creative energy available in the rich tapestry of social behaviour where every moment is a new opportunity to invent and re-invent social reality. If we behave differently the world is a different place. If, in the process, we are being true to ourselves such changes are likely to be lasting.

What we do need to remember in this context, however, is that, in the quest to build networks of relationships that will encourage self-realisation, it is likely that the way we behave towards one another in practice will be *the* central theme in creating an atmosphere of mutual understanding and respect. The more we can discuss what we are doing and why, in an accepting environment, the more likely it is that we will have to face ourselves. And it is through learning and accepting who we are that lasting growth occurs. The same is true for relationships *between* communities.

The fact that much of our social behaviour is governed by roles is one reason why it is so difficult to imagine the kind of community in which such networks of relationships might exist. For most of the time we are not being *ourselves* relating to other people who are being *themselves*. We are acting out ways of relating to one another that may have their roots in social reality many centuries ago. Everything we do and say has a social context that exists outside ourselves. The way people behave towards one another is itself buttressed by the social structure and value system of the wider society of which they are a part. These, in turn, have a momentum and a logic of their own that can sometimes make it difficult for people to behave in the ways they wish, or to change conduct that is no longer beneficial. The expectations of one's culture become deeply internalised, a circumstance that is nowhere more apparent than in our ongoing attempts to explore what it means to be male and female; 'traditional' views of how the sexes should relate always get in the way.

Eating People is wrong – the plane of social structure
This social 'skeleton', which exists before we are born and will still be there when we are long gone, plays a vital function in any grouping of people. Even the most carefully selected group of individuals on our desert island will have disputes to resolve. Probably they will sit down on the beach, with all the time in the world, and talk through their differences, trying to find outcomes that will suit the circumstances, the personalities involved and the relationships that exist between them. If they don't succeed, or don't think it worth the effort, they will end up knocking each other about until one or more has established dominance over the rest.

Above a certain number of players, however, this essentially people-centred approach becomes impracticable. And yet, unless mechanisms exist to find satisfactory solutions to grievances, there will be a tendency for quarrels to simmer on, to degenerate into brawling or for the social fabric to disintegrate into factions which may, ultimately, lead to civil war. The ways in which small-scale societies avoid this danger is through face-to-face exchanges that are ritualised and which usually involve the intervention or mediation of a third party whose 'authority' is accepted by both parties. 'Rules' are not automatically obeyed and any culture needs sanctions, both to ensure they are accepted and to deal with transgressors. Already, we can see the emergence of quite complex social imperatives that are more or less independent of the real people who form the society at any particular time; they are 'structural' in nature.

At this stage of development, however, the ways in which disagreements are handled are verbal and they tend to be invested in actual individuals. Thus, the way in which an elder in one generation interprets the rules may be very different from their successor's way; that difference reflects who they are as people rather than any objective change in circumstances. Solomon was seen as a wise judge because of his personal qualities rather than because he administered a system of wise laws.

The problems with this approach are obvious. It will tend to be less and less effective the greater the number of people involved and the greater the distances that separate them – which is why empires have always depended on force to sustain themselves. Until comparatively recent times in Europe (and still in many places in the world) areas existed that were 'beyond the law' because the word of the law couldn't be heard there or because it could be safely ignored. Writing offered a partial solution, in that laws could be literally inscribed on tablets of stone and could be offered as a visible extension of a distant authority. Documents also took on an significance of their own, providing evidence of rights and allowing the panoply of justice to be transmitted effectively over time. It was one thing to talk about executing Charles I; it was quite another to sign the actual death warrant – and the consequences for the regicides only a few years later showed they were right to be frightened of putting pen to paper!

Printing allowed organisation on a massive scale for the first time. Not only could rules be churned out remorselessly, they could be distributed through increasingly sophisticated communication systems that depended on consistent and standardised responses that were themselves governed by rules. Technological innovation fuelled the rush to conformity so that, with the advent of the railways, time itself was tied down and divided into ever more precise units. Computers, satellites and the information superhighway offer us the potential to regiment the whole globe under one, universal authority.

One thing is clear from this process. The social structure is more important today than it has ever been and is set fair to become even more dominant. A consequence of this trend has been for authority figures to assume 'roles' and to act according to the logic of the rules they personify even when those rules are patently wrong. Structures have become institutional in the sense that they actively discourage *individual* interpretation. Elaborate appeal systems are created to test the validity of judgements; the criterion used is not whether a particular decision is 'right' (did it achieve what it intended? – the Solomon test) but whether it is methodologically sound and conforms to the body of precedent that has been established over time. In other words the system has taken on a logic of its own which may or may not accord with the real world outside. It is all a far cry from our little group on the beach!

Power lies in these institutions, and it is the power to ignore individuals, treating their problems not as uniquely theirs – with solutions that will also be uniquely theirs – but as an abstraction that can be measured and dealt with by applying a standard set of responses that will be the same in every situation. That is why the individual so frequently comes away from encounters with the institutional world with the very feelings of anger, bitterness and of not having been listened to that any person-orientated media-tion system should be striving to minimise. Even when one 'wins', one is left feeling that natural justice has not been done. And, when one loses, the resentment can last a lifetime. That is why the creation of communities within which personal self-realisation can be a reality seems such a distant possibility; it is the institutional world that now determines the norms of inter-personal behaviour in most key areas, not ourselves. In fact, institutions thrive on de-fining social behaviour, from prescribing 'uniforms' to differentiat-ing grades or functions, to the introduction of specialised languages and rituals. Put simply, the level of social structure has come to dominate that of social behaviour.

The essentially adversarial nature of the institutional world is evident in the way that one institution breeds another; action and reaction. Campaigns against the worst excesses of a particular busi-ness or state bureaucracy invariably lead to 'protective' legislation and the setting up of a 'watchdog'; the bigger the organisation that needs to be controlled the bigger the monitoring agency has to become to be able to challenge it effectively. In seeking redress, individuals have to engage a legalistic process that interprets their complaint according to a fixed agenda that can make little sense to the layperson. They become of interest only in so far as they repre-sent part of the battleground on which institutional forces struggle for dominance. Once again their individuality – and the unique-ness of *their* problem – is denied.

This tendency to produce and exaggerate conflict is present in all institutionally managed situations, from the trade union representing its member at a disciplinary hearing to an appeal against the construction of a new motorway. The human impulse to minimise conflict at the personal level through dialogue is absent because the debate is not about real human relationships. Principles and precedents are the issues being discussed, and individual needs are frequently trampled underfoot in the set pieces that are needed to resolve them. As the power of the institutional world grows, so the adversarial approach to problem solving spreads to more and more of everyday life. Indeed, it can be argued that, in an institutional world, the state can respond only by creating more laws because it has no effective way of entering a dialogue with its constituency, the people. Once *human* debate has ceased, only rules and their interpretation remain.

In an age that prides itself on championing the rights of the individual, and on being a mature democracy in which all are equal before the ballot box, such conclusions should be worryingly at odds with received wisdom. Are we not 'free' in a way that has never been achieved in history before? Do we not have a system of social justice that should be the envy of the world? The reality is that, just as there is often a divergence between how real people behave and the roles that the social structure ascribes (and such departures are often labelled 'deviant' or 'delinquent'), so there are often major discrepancies between the values a society holds dear and the way in which the social structure (which is the vehicle by which values are interpreted in practice) actually responds.

It's what you believe that counts – a first look at values
Some of the most striking examples of this mismatch between what institutions – and the people who work in them – believe they are about (their core values or principles) and what they do in practice can be found in the health and welfare system. Most people who work in the caring professions would claim that they are there to help other human beings to get better or to overcome disabilities and lead valued lives. Time and again, however, the experiences of people who receive these services are quite the opposite, leaving them feeling devalued and de-skilled. Prolonged exposure to such care can lead to institutionalisation, with the

individual rendered incapable of making independent decisions about even the simplest life choices, such as what clothes to wear.

A moment's thought will suggest why this should be. If I suffer an appendicitis it is a major, and I hope unique, life event. It will have great significance for me, far beyond the physical manifestations, temporarily defining who I am for both myself and others. From the hospital's point of view, however, I am only one of many hundred cases of suspected appendicitis that they will treat each year. The chance that I will have anything positive to contribute to the process they about to undertake is very small. Immediately, there is a mismatch of expectations, with me needing to be treated as uniquely myself, experiencing something new and overwhelming, and the hospital wanting only to get on with a well-rehearsed routine in which 'I' am incidental. When you add to that fact the sheer impossibility of nursing staff forming meaningful relationships with each of the patients who pass through their wards, we can begin to understand why hospitals (and the same thing happens in any institutional setting) cease to treat people as people.

The simple fact is that caring institutions are like factories and are created to process a certain number of objects (who happen to be human beings) through a preordained set of operations – getting up and going to bed, bathing, toileting, eating, etc. – in the most efficient and cost-effective manner. That inevitably means fitting people (the objects) into routines, and the best routines are the ones that process people the quickest – because that means less staff time at less cost. As long as people fit into those routines everything is fine and the carers can feel that they are looking after everyone's needs and doing a demanding job in a professional manner.

If someone objects, or gets in the way of the routine, workers face a dilemma. Their training and experience have shown them that the only way to cope (and meet everyone's need to be bathed, etc.) depends on the smooth running of the system. On the other hand, a residual identification with the patient leads to an initial wish to find a way of changing things to accommodate their demands. But adapting the system is, at best, extremely difficult and it is easier (and it gets easier and easier with the passage of time) to ignore the individual or define them as 'difficult'.

The problem then is to get them back into the routine; once that stage is reached, abuse is already a silent partner in the transaction. It may start off with shouting at the individual and progress to manhandling them, tying them up or beating them. More subtle techniques may be employed, such as sedation (always in the patient's best interests), but the effect is the same. It is frightening how quickly people living *and working* in institutions come to accept such a reduced view of life's possibilities. These are not bad people, just individuals who have come to accept rhythms and routines that are primarily geared to the smooth running of the institution and not the recognition of patients as people with individual needs. The social structure has taken over and staff are as institutionalised as the patients they care for. No one notices that the institution has lost sight of its original *raison d'être* and is working to an entirely different agenda. The staff still believe they are operating a principled service, and the fact that a new mission statement emerges from the board room each year only confirms this view. Meanwhile, on the shop floor, nothing changes.

The tendency for practice to become detached from principle is one of the features of the institutional world and emphasises the dangers inherent in our mass society where, as we have already seen, it is institutions that increasingly define what is possible at the level of social behaviour. This separation between values and what the social structure delivers in practice is hardly a new phenomenon. All societies have sought to justify their hierarchical structures through value systems that have little to do with what is actually going on. Invoking those values, however, allows the key players to pursue their own self-interest, firm in the belief that it is synonymous with that of society as a whole. In the end, the tensions between rhetoric and reality become so strained that the counterfeit is revealed for what it is. The struggle for power then becomes an open battle between rival factions, each seeking support through the witness of its particular ideology. Victory eventually goes to the side whose stated principles are most in accord with the times. Initially, at least, those principles may well reform practice but, inevitably, there is a falling away. And the process starts all over again.

Or it has up till now. The situation today is unique both in terms of the scale of the institutional world and in terms of its all-embracing nature. There is nowhere left to hide. Just beyond the circle of our daily lives the big battalions are camped and, when things go wrong or we attempt to step outside our preordained routine, we come up against the hard, unyielding face of their power, the power to ignore us and ride roughshod over our uniqueness by reducing us to a category or a number and dealing with us accordingly. It is an essentially dehumanising animal, a hydra-headed monster that is inherently difficult to slay because we have taken it into ourselves. In a very real sense, we have to destroy the institution within ourselves, and that means confronting our dependence on it. *For we are all institutionalised to a greater or lesser extent.*

It is a start to recognise that we have one major asset. As suggested above, we, and no one else, are ultimately responsible for *our* behaviour. If we see that it is in our interests to change the nature of the relationships we have with the people around us, we can set to and do something about it. What should now also be evident is that we underestimate the power of the institutional world, and the influence it has on our behaviour, at our peril. To avoid that danger we must understand where it is coming from and be able to challenge where it is weakest – in its inability to deliver the values it professes to espouse. We must also attempt to summon an alternative vision that will offer a sense of direction as we begin to move off into the unknown.

Before doing that, however, it is worth reflecting on the consequences if we don't succeed. One possibility is that the institutional world will continue to grow and become more powerful and all-pervasive. Ultimately, it will dominate the globe in a way that will makes today's social structure seem positively humanitarian. As we have seen, institutions are particularly prone to moral blindness, and institutionalisation on this scale would almost certainly be accompanied by a complete denial of individuality. Lewis Mumford's 'megamachine'[6] – a society run as a single institution – can be likened to running a car. Although a complex piece of machinery, a car is essentially easy to operate. Where the driver wishes to go, however, is of no concern to the megamachine, which is intent only on performing its prescribed tasks, even as it heads over a cliff or into a brick wall. It is not hard to see how such a society, inherently repressive and authoritarian, could quickly become a nightmare. 'If you want a picture of the future, imagine a boot stamping on a human face – for ever.'[7]

Another scenario is hardly less apocalyptic. As we become ever more partial and specialised beings so we will understand ourselves less and less. Rather than the world becoming increasingly routinised, it will become prone to virulent outbursts of irrational behaviour as our unconscious rebels against the straitjacket of our existence and the lack of self-esteem it offers us. In such circumstances, we are likely to respond to threats with the armoury of the unconscious: denial ('it's not really happening' or 'it's not as bad as they make out'), projection ('it's the fault of the Russians' or 'them'), sublimation ('I'm far too busy keeping a roof over my head to be bothered with that'), etc. We will ignore the violence being done to others until it suddenly bursts around our own head; and then, of course, as we look around for help, we'll find everyone else busily ignoring what is happening to us. Alternatively, we will be swept up in the apparent justice of what is being done to others because we are projecting onto them the bits of ourselves we don't accept, or won't acknowledge.

Once that kind of blind energy is let loose in a society it has to take its course. The individual becomes submerged in the psychology of the mass and there are no checks other than the distorted logic of the collective message. Few individuals have the moral courage to stand against such a flood tide because the conditions in which moral beings are created and sustained (an environment in which individuality is valued and daily affirmed) have long since disappeared. The choice is usually between complete submission and bringing the hatred down on oneself.

It is one of the myths of our age that reason can find solutions to everything. In gaining mastery of the material world we think we have gained mastery of ourselves. Even a cursory glance at the world we have created, with its injustice, cruelty and simple disregard, must give the lie to that convenient fiction. We are moved by forces we barely acknowledge, let alone understand.

In the mythology of earlier times, these forces were called mana, or spirits, demons and gods. They are as active today as they ever were. If they conform to our wishes, we call them happy hunches or impulses and pat ourselves on the back for being smart fellows. If they go against us, then we say that it was just bad luck, or that certain people are against us, or that the cause of our misfortunes must be pathological. The one thing we refuse to admit is that we are dependent upon 'powers' that are beyond our control.[8]

It is indefensible hubris to believe that reason alone can triumph. If we do not learn to live with those 'powers' – and that means gaining a deeper understanding of ourselves and the true nature of the universe we inhabit – they will destroy us.

2

The Unholy Alliance

There are more things in heaven and earth, Horatio
Than are dreamt of in your philosophy.

SHAKESPEARE

A death, no matter how peripheral to our circle of family and friends, invariably gives pause for thought. For a while we find ourselves considering what is important in life, perhaps even making fresh commitments to ourselves and others. The fact that very few of these resolutions find their way into daily practice – determined by priorities that seem outside our control – should not be taken as evidence that we don't care. Each of us has a set of values (including some sense of the purpose of existence) that form a more or less conscious point of reference for our lives.

Such statements of importance are rarely consistent, either within individuals or within society, and from the tensions generated between them come our moral dilemmas. Values, and how they are expressed, can change dramatically over time. The world seen through the eyes of a medieval knight would have been very different to that of a courtier at Versailles or, yet again, an industrialist in Victorian England. Basic human needs, however, do not change: our hierarchy is as relevant to a cave dweller as to an astronaut on the Moon. How is it that, from the same starting point, such different perspectives can evolve, perspectives that can lead one people to attempt to conquer and even exterminate another?

A cat or a dog experiences the world in a direct, visceral way. The sights, sounds, smells and tastes of the garden are felt immediately and deeply, producing a largely stereotyped set of responses. The myth of the 'Fall' suggests that we once existed in the same direct way, our inner selves and our environment merging into a

single state of being. The eating of the fruit of the tree of knowledge symbolises our attempt to detach ourselves from our surroundings, to stand outside them and, in time, to seek mastery over
them. In the process, we have turned our backs on paradise and
must wander the earth searching for meaning to replace the
certainty of being at one with creation. We have become concep-
tual beings, filtering reality through constructs that we find 'out
there' in society and which we absorb, largely uncritically, as we
grow and develop.

For the medieval warrior reality was framed by the threat of
sudden violence. The important thing was how a man – for
women hardly counted in this world view – presented himself. A
'hero' could win kingdoms through fortitude and bravery, or ride
into legend on stories of how he lived and died. By contrast, life at
the Court of Louis XIV was characterised by cunning, man-
oeuvring and the manipulation of events, all carried on behind a
facade of courtly behaviour. The ideal was long-term advancement
through the cultivation of elegant and apparently effortless
manners. For the Victorian industrialist the market was everything
and, once again, the rules of the game were understood, if not spelt
out. An individual's stature depended on his astuteness and nerve
within the limits of commercial practice; one might treat one's
workers like a sub-species but to be caught cheating in a business
deal meant ruin. Each of these ways of facing the world can be
summarised by simple statements of belief, 'right is might', 'the
divine right of kings', 'my word is my bond', which both legiti-
mised behaviour and gave it shape. As we have seen, real people
didn't always, or even often, live by these convictions but they
nevertheless gave an overall sense of direction and purpose and
provided meaning and credibility to what they actually did.

We cope with being the least instinctual of living beings by
creating and internalising such 'mind maps' to explain what is
happening and how best to respond to circumstances. From the
prevailing values we construct the rich pageant of social behaviour
as well as a social structure that both gives it weight and ensures
compliance. That is why from time to time it is worth checking
out the values that actually underpin our mind maps. Confusion

invariably exaggerates injustice, conflict and disharmony simply because we don't take the time or trouble to ask whether behaviours we don't like are calculatedly anti-social (in terms of our own assumptions about what is acceptable) or the result of a slightly different way of looking at life.

Although we have a strong tendency to stick with tried and tested views of the world, even when the evidence suggests they are no longer working, we are, in practice, continually simplifying and overhauling our concepts in an attempt to maintain a sense of unity and purpose. Very occasionally, we are capable of a 'Road to Damascus' conversion, when one mind map is completely substituted for another. In a changing world, especially, we are also superficially more amenable to new ideas – which, in our age and the particular values it espouses, become a product like everything else, to be marketed and consumed.

Despite this capacity for abstract thought, no blueprint so far advanced has been universally applicable. 'We can approach the sacred mountain of Truth from many sides. What, in the natural limitation of our faculties, we cannot do is to see all sides at once. There is only one point from which we could do that, and no one has yet been there.'[1] Each mind map has its inherent limitations that, over time, leads to its being replaced by another, more appropriate if equally limited, credo. The range, flexibility and subtlety of language will be one factor in determining how versatile our mental pictures of reality can be. It was George Orwell's greatest nightmare that a world of Newspeak would so restrict and impoverish language as to make thinking outside the accepted norm impossible, because the words needed to express the ideas would have ceased to exist or been robbed of any substantial meaning.

This tendency to over-simplify or to leave out bits of the overall picture has many consequences. It can make it difficult to express some concepts that, at other times or in other cultures, were easier to convey. The Eskimo have many words for 'snow' and the state of 'snowing', allowing for a rich variety of responses to conditions that can spell life or death. By contrast, our perception of a not infrequent occurrence is limited to the crude distinction between snow itself and slush, with flurries, squalls and blizzards to give some indication of severity. In our competitive world it is equally difficult for many to find, let alone express, the gentler side of their natures. 'When the going gets tough, the tough get going' has become almost a creed, substituting doing for thinking or feeling as the highest attribute we can aspire to.

The fact that certain aspects of ourselves are not being expressed does not mean they don't exist. On a psychological level, outlets have to be found and, just as a stream carves out a subterranean passage under a mountain, our deepest, unmet needs can emerge in surprising ways. Science has been so successful in transforming our lives that, for many people, it has become a religion in all but name. Its authority is as absolute as the Church of the Middle Ages and its institutions similarly define and legitimise areas of doctrinal interest and reward the great and the good. The impact of science on everyday life also finds echoes in the hold that the Church once had over people's minds. Unable to grasp the subtler points of theology, the masses recited homilies and looked to the example of the saints. The litanies of science are just as homely (Newton and the apple) and the great figures of science provide models of how truth is to be pursued. Secure in this mind map – with the sense of belonging and prestige it provides – the prophets of the scientific method are able to mock the 'faith' of churchgoers and, if circumstances permitted, would undoubtedly throw them into outer darkness as heretics, happy that they were acting for the best.

That is another consequence of our inability to tie reality down. The bits we fail to acknowledge have a habit of coming back to haunt us when we finally have to confront them in others. The stereotypes we have say more about ourselves than the people they describe. Thus, the amazingly diverse peoples of Africa are often portrayed as universally lazy, feckless and sexually promiscuous, all aspects of our own culture that we are less than comfortable with. *Our* problems create the barrier but we *perceive* them as being inherent in the racial characteristics of another people. The unspoken implication is that, if only *they* would change and be more like us, we could get along fine. It allows us to retain a sense of superiority without having to delve too deeply into our own assumptions. We do not have to face the inadequacies in ourselves and we are less complete as a result. Such undercurrents may not start wars and pogroms but, once under way, they can ignite the arid, suppressed regions of our subconscious with frightening consequences.

To understand a people or a culture, then, a good starting point is to look at the values it professes, both to see how practice measures up and to get insights into what are likely to be blind spots. We live in a pluralistic age with an 'anything goes' attitude, but beneath that tolerance – some would say complacency – there clearly lie some fundamental concepts about what a decent society should be about. Most people in the Western world would subscribe to notions of democracy, economic growth, success through individual opportunity and reward, social justice, family and a sense of community. What do these concepts mean and how coherent are they?

'If a man will risk his life to protect his family and homeland, surely he deserves a voice in matters of war and peace and a say in how the City should be governed.' So argued Cleisthenes around 500 BC, in laying the foundations for the democratic experiment in Greece. '*Demos Kratos*' equals 'People Power'. In our so-called mature democracies that has come to mean participation in a periodic jamboree in which the electorate are invited to cast their votes in favour of a vague agenda that has little binding power on a party once in office. Presentation and the sound bite count for everything – 'read my lips, no new taxes'[2] – and issues are massaged into and out of existence. Politicians are groomed and coached to fulfil the leader role; fit guardians for the strong and free. The political scene, especially at the international level, has become a long-running soap opera with regular flurries of excitement as key figures come and go against a backdrop of scandals, coups, wars and other threats to world order.

Public opinion may still terrify politicians, but only because it is a fickle, populist tide of feeling that has more to do with the Roman amphitheatre than the ultimate ideal of Greek democracy, which was to ensure collective responsibility for decisions that affected the well-being of all. In practice, governments that do not even command the support of the majority can exercise an essentially autocratic rule for years on end. When the opportunity to elect a new administration does arise, a significant minority vote with their feet and stay at home. To the Athenian, the citizen who failed to participate fully in the business of the city was useless; it was an abnegation of that individual's responsibility. The lack of involvement in political debate at either local or national level, so apparent in today's democracies, would have been incomprehensible two and a half thousand years ago. Our 'mind maps' have no words for that sense of participation in, and ownership of, everyday affairs. So much for progress!

One of the most striking differences between the ancient Greeks and ourselves is the sheer number of 'citizens' who potentially need to be involved in decision-making. Athenians could literally stand up in the market place and address their fellow citizens; they could make themselves heard. Despite our technological wizardry, mass societies have failed to find mechanisms that allow that direct, personal intervention and are immeasurably the poorer for it. In fact, the opportunities for real participation are limited to taking on the role of politician (and what a strange role it is, flitting around the country – if not the world – talking, endlessly talking, to complete strangers, and all the while divorced from the daily realities that affect the rest of us) or reacting through protest. The alternative, to hand responsibility to politicians in the belief that they somehow represent their constituents, is the worst of all possible self-deceptions, if only because every political act is increasingly influenced by the economy and the continuing search for economic growth.

But, surely, economic growth is good for us? Is it not the way in which wealth is created, jobs secured and services improved? Up to a point that is true, but to have more of something and have it better made doesn't necessarily equate with a better quality of life; the impact of car ownership on our environment and the fact that we can't now live without personal transport despite the lengthening queues, the impact on our health, etc., etc., shows that increasing wealth can make life *worse* for everyone.

There is an even more fundamental question here – what is the goal of economic growth? Assuming that we *do* want all humankind ultimately to share in the benefits of a growing world economy, what kind of existence will it be? Do we want everyone on the planet to own a car? If so, what kind? Will we be satisfied with a basic Ford or should we all aspire to a Mercedes or a Rolls Royce? Should it be replaced with a new model every year or every five years? What will be the impact on life when whole communities are swept away to build and re-build the road networks needed to sustain the billions of vehicles there will be? And where are the raw materials to build them (and the other consumer durables we might have a right to expect)? And the power to keep them moving?

In short, it isn't going to happen. And yet a small proportion of the world's population continues to consume at an ever more frenetic rate, secure in the belief that economic growth is good for you. Even after the legacy of Chernobyl, we continue to build and run nuclear power stations as an insurance policy against the day the oil runs out, hoping against hope that new, greener technologies will come along and allow us to continue with our energy-profligate lifestyles unchanged. We create ever larger economic units to smooth the way for venture capital to work its wonders on the balance sheet without questioning the impact on the communities that are sacrificed in the process. We harness ourselves to the plough of work without asking why people feel lost when that purpose is removed. All life's problems are temporary aberrations, necessary for continuing expansion – even if the solution of one problem is the creation of two bigger ones; they too will be overcome in time.

It is the mind set of the addict who knows that the habit is killing them but can't break out of the vicious circle that is creating an ever greater craving. Money, and the peculiar value we attach to it as a culture, provides a first clue to the nature of the disease. Firstly, there is an increasing tendency to want to put a price on everything. Money is thereby elevated to the highest of all values because, by definition, all other values can be expressed in terms of it. In the process money gains a spurious objectivity that makes arguing against it extremely difficult. It has become a closed system, an end in itself, that allows us, for example, to go on consuming raw materials as if they were going to last for ever. The

Earth has become, in effect, one vast supermarket store from which we can purchase whatever takes our fancy, secure in the knowledge that it's someone else's responsibility (it's always someone else's responsibility) to keep stacking the shelves.

Secondly, money transactions suspend individual morality – the basis on which we interact one with the other in practice. Although the market place provides consumers with great power by offering a range of choices for any given outlay, it also limits our responsibility in the transaction to getting the best price we can. The conditions in the factory that produced the goods are not our concern (even though we pay for unsafe or unhygienic conditions through increased healthcare contributions): nor are the hazards found in the Latin American mines that produce the metals the factory uses, even though they result in a life expectancy for miners of less than forty years. When we make a purchase we are involved in a deal of the moment, and such considerations exist outside that moment. They are, once again, someone else's responsibility. Such a defence would hardly succeed if the goods had been stolen. At best, we would lose what we had acquired, at worst we would be charged with being accessories after the fact. The value we place on the forces of the market place, however, allows us to get off scot-free. We are encouraged to hold a partial vision and, in the process, have become less than human.

The stock markets of the world are prime examples of the limited perspective that money instils. Prices go up and down in response to how individuals (only those individuals with sufficient money, of course) perceive their maximum – usually short-term – advantage. And that advantage is defined solely in terms of financial gain; make money and you've done well, lose it and you're an also-ran. Because of the distorted view this simplified approach to life produces, unpredictable events on the other side of the world can have a dramatic impact on plans that may have taken years to unfold. Individuals, communities and even nations can be ruined, as it were, by a single roll of the dice. Billions of dollars a day are 'invested' in this way, instantaneously switched around the global economy with no thought whatever to consequences. It is a game, and winning is everything Nothing demonstrates more clearly how money has become detached from the needs of real, flesh and blood human beings.

Yet, money is a human construct and one that was originally marginal to social interaction. The bulk of human exchange was conducted through mutual aid based on family and kinship. Barter existed to supply those wants that the group itself was unable to meet. Money, as a token of exchange, developed only as the boundaries of trade were pushed back with improved and secure transportation systems and a widening range of goods to market – in other words, as states became larger and more organised. All three forms of exchange exist today, although the money economy is the only one officially recognised. In terms of Gross Domestic Product, that all important assessment of how an economy is doing, the other two don't even feature.

So what does GDP measure? The simple answer is that, in so far as it tells us anything, GDP is a measure of the growth (or lack of growth) in the money economy. On this basis, the more car crashes there are, and subsequent repair work, the healthier GDP looks. Indeed, it is possible to envisage a society in which the economy – as measured by GDP – is steadily growing through the successful manufacture and export of endlessly changing designer novelties, while the quality of life is plummeting, pollution and unemployment soaring, roads – apart from those necessary to sustain economic activity – crumbling, schools becoming cold, uninspiring gaols and a health service overwhelmed by demand for basic care. Such a scenario is commonplace in the Third World and may not be so far off in our own.

One of the main justifications for a growing economy is the need to create jobs – the way in which individuals in our culture seek opportunities and rewards. Unfortunately, competition in the market place leads inevitably to job *losses* because the need to remain competitive implies increasing productivity, which is achieved by reducing the labour element on the balance sheet by improved technology (or lower wages, which are harder to sustain politically in our reward-led society). The economy has to grow, therefore, to create the additional jobs needed to replace those that are being squeezed out elsewhere. Politicians of all persuasions see continuing growth as their salvation because they are so terrified of the consequences of the economy stalling and beginning to contract.

With competition for markets both at home and abroad becoming ever fiercer, the nation – or group of nations – that gets it right in terms of technological productivity, organisational flexibility and 'realistic' wage levels – in other words, maximises profitability – will emerge triumphant. That is now the prize to which political parties of all persuasions are committed. Creating a model welfare state or increased democratisation are yesterday's goals, to be indulged in only if they can be afforded along the way.

Rising unemployment, company closures, bankruptcies, redundancies and house repossessions contribute to the long litany of recession and slump. But what really happens when an industry like coal mining collapses, throwing whole neighbourhoods out of work, or a major employer relocates leaving little prospect of employment to those left behind? The people remain the same, their skills are still intact, the area hasn't changed. What *has* changed is that the amount of money available locally has been drastically reduced and, as a consequence, people can no longer pay for goods and services they once took for granted. More importantly, there appears to be no alternative to money as the medium through which needs are met. Our lives have become defined by money, and that is nowhere more apparent than when we haven't got any.

Of course, there is still plenty of money whizzing round and round the world but it has ceased to stop off where we live. It is going to fund *profitable* enterprises elsewhere and, all of a sudden, another location on the other side of the globe finds itself awash with cash, able to afford the ephemeral prestige of the latest fashions – all created by the advertising agencies to separate us from what money we have got as quickly as possible so that it will move on and generate yet more profit.

It is a merry-go-round that, looked at from the point of view of meeting even the most basic of human needs, makes little sense and whose only moral justification is an endless and increasingly frenetic pursuit of the greatest return on the basis that one day (some day, never), when the system is perfected, everyone will be better off. In the meantime real people are hurting badly. Success is shown to be illusory and the rewards are never sufficient. Families and communities disintegrate under the twin pressures of over-work and the need to keep moving on in pursuit of the next promotion – or they crash into the abyss of having no work and no hope of ever moving on.

Nothing is what it seems – or so it would appear
In terms of the values we would like to believe we are living by, then, we are clearly not doing very well. Is it that we are not trying hard enough or are we being forced – or perhaps choosing – to live according to an entirely different agenda, one that we would rather not acknowledge because it doesn't square with the image we have of ourselves? What *are* the major influences on our lives?

Thus far, we have seen how institutions have come to define the limits of what we can do as individuals and how they take on a logic of their own that enables them to act in ways that are at complete variance with their stated aims. We have also seen how the money economy is now so pervasive that anything can be valued in its terms and then exchanged in a deal of the moment that places no responsibility on the parties in the transaction beyond getting the best price possible. These forces overlap and reinforce one another. Without organisations based on rules that are consistent over time and place, the money economy could not have become so all-embracing, nor could its peculiar moral blind-ness have been sustained. Without the money economy providing a universal medium of exchange, organisations could not have grown to a point where they span the globe. It has been a remark-able achievement and one that has produced a truly world order for the first time in human history. But it has been bought at a terrible price, nothing less than the dehumanising of life through the denial of your uniqueness and mine.

What has happened to allow this state of affairs to occur? Determining cause and effect is always dangerous because tinker-ing with any bit of a system as complex as human society inevita-bly has consequences on all other parts which, in turn, reverberate backwards and forwards until a new, temporary stability has been achieved. I think it is nevertheless possible to distinguish a gradual and continuing shift in the way we view the world (the level of values), a change that has not happened overnight and is not yet complete, but one that has underpinned the phenomenal growth in the money economy and institutional world to produce a social order that no longer cares about people.

Put simply, there is a juxtaposition between a spiritual world view and a material world view. In the West, we have notionally moved from a culture dominated by the former to one that is at best confused. Without exception, the spiritual world view places value on each and every human being (by virtue of the fact that they have been created by a God who is concerned for them) and on the way we behave towards each other. The material world view doesn't imply that individuals aren't important – humanism, after all, has a fundamental commitment to human welfare – but it does allow the possibility. Values, as we have seen, tend to be elastic, especially in the institutional world that has come to dominate the social agenda. If they can be stretched a little bit further and still embrace the behaviour in question they will be. Where there is a chance, therefore, that other people can be ignored without the importance of individual life being called into question, there is a certainty that somewhere along the line it will be tested out. And because we realise that the rubber band was overstretched only when it actually breaks, we may not even be aware of what is happening.

At this stage it is important to make the distinction between the spiritual world view and the worldly consequences that are often associated with it. We are only too familiar with the catalogue of war, murder and atrocity that has been the legacy of the Christian Church down the centuries, but the two are not necessarily causally linked. 'European–North American history, *in spite of the conversion to the church* [my italics], is a history of conquest, pride, greed; our highest values are: to be stronger than others, to be victorious, to conquer others and exploit them. These values coincide with our ideal of "manliness": only the one who can fight and conquer is a man; anyone who is not strong in the use of force is weak, i.e. "unmanly".'[3] In other words the pagan celebration of the hero remained the central dynamic of Western history, despite the apparent importance of a more overtly spiritual world view, and was always the prime mover in the religious institutions that were supposed to promote that ethic: yet another example of the social structure acting in complete opposition to the values it professes to embody! Jung observed something remarkably similar: 'The great events of our world as planned and executed by man do not breathe the spirit of Christianity but rather of unadorned paganism. These things originate in a psychic condition that has remained archaic and has not been even remotely touched by Christianity.'[4] We now live in a phase of Western history where the impact of the spiritual world view, with its emphasis on the individual, is at its lowest ebb. It is perhaps not surprising, therefore, that we find ourselves in a culture that is as rapacious and as unconcerned about the individual as at any time in the world's history.

In part, this diminishing of the individual is also an inevitable consequence of the vast increase in the world's population coupled to the fact that we know what is going on all over the globe. Disasters, both natural and human-made, flash across our TV screens, sweeping away hundreds if not thousands of unnamed and unknown individuals in their wake. And we carry on eating our meals as if nothing had happened. We really have no choice, but it shows how hard it is to hold on to an idea that every human being is valuable.

It couldn't have happened, however, without the rise of a mind map that presented a radical alternative to the spiritual world view. For most of human history, the realities of life, however grim and violent they may have been in practice, were tempered by a sense of life's purpose that was other-worldly (for example, this life as a preparation for the next). This recognition, and its evolving interpretation, of a non-material dimension had a profound effect on the way the natural world was viewed. Initially it was seen as being essentially mysterious, often capricious and frightening, a power that had to be appeased: imagery that persists in hymns and animistic traditions. As life became more settled, so the vision became more tranquil, with the land depicted as 'God's garden' to be husbanded and celebrated as God's gift to humanity. Although these metaphors were still commonplace at the time Galileo was summoned before the Inquisition in 1616 for demonstrating that the Sun, and not the Earth, lay at the centre of the universe, they were already losing authority. His public recantation emphasised the temporal power still wielded by the Holy See but already its monopoly over what people believed was far from absolute; dogma was being replaced by a more open-minded scepticism.

Over the next century and a half there was a gradual transition to a belief in the divinity of Nature and thence to a conviction that the world was rational and explicable. Descartes, the first modern thinker to present a coherent philosophical system that was not founded on traditional theology, embodied the ambiguities of this shift in perception. Although he believed absolutely in the pre-eminence of human reason, he nevertheless acknowledged the existence of a 'designer' who had provided humanity with its rationality and who, therefore, effectively underwrote its authority as a means to understand the world. It remained only for human reason to propose that 'God is a hypothesis which is quite un-necessary'[5] for humankind, through science, to begin its conquest of nature. The pagan in us had found a way of viewing the world that gave us carte blanche to plunder where we pleased. And that includes the use and abuse of our fellow human beings.

It is important to emphasise that the scientific method, or even 'rationality', is not the villain of the piece. The list of benefits from scientific discovery is unending. That is why the mind map provided by science appears so plausible. What other thought system in the history of the world has contributed so much to the overall welfare of humanity? Its very success is the reason why it has become so dangerous. The nature of reality being what it is, the world revealed by science cannot be the whole story. We might have got a better view of the 'sacred mountain of Truth' than ever before but it is still only the view from one side. In some ways, it is easy to see where the problem lies. What is less straightforward is achieving the vision that will provide a more balanced, or 'holistic' vantage point without losing the understanding that science offers from its own unique perspective.[6]

In their search for artificial intelligence, computer engineers can now design programmes that act as 'experts' in certain clearly defined areas of activity. The diagnosis of blood disorders, for example, can be done as effectively and more reliably by computers than by a human specialist with years of training, because the information provided by batteries of tests can be processed with a single-minded consistency not attainable by the human brain. If you ask the same computer what a patient is, however, or whether they normally prefer to 'live' or to 'die', the limits to their 'intelligence' become clear. Our humanity depends on our ability to bring together any number of apparently unrelated facts and make a human decision about their significance. Implicit in that process is the application of certain value frameworks that grow out of our conception of what being human is all about. Part of that heritage is that we are creatures who need to relate one with another in a human way, which means being able to put ourselves in the position of the 'other' and interact through a process of give and take (not, you will note, through maintaining our emotional distance by a carefully cultivated objectivity).

The increasingly specialised nature of our interactions through the medium of roles is eroding our innate humanness, firstly because the value frameworks that underpin the roles we adopt are institutional in nature (and determined, through the need to maintain profitability, by the money economy) and, secondly, because their specialised nature means that we no longer 'see' the other or understand his or her (equally specialised) perspective. The irony is that, far from being able to construct computers that can think like humans, we are in danger of creating humans that think like computers.

There are many phenomena that can be understood only as being greater than the sum of the parts that compose them. For example, the analytical approach of science can offer insights into the human brain, indicating the location of certain functions, but from that information it would be impossible to predict – the test of a scientific hypothesis, that it was capable of something as abstract as thinking – the very thing that makes scientific endeavour possible. As humans, *we* make the jump between knowing of the existence of a lump of grey matter and our capacity to think because we can move between one level of meaning and another without difficulty.

We intuitively accept that there are higher- and lower-order functions and that it is impossible to understand the former purely through a study of the latter. A society is *not* just the sum of the individuals who make it up any more than we, as individuals, are the aggregate of our bodily parts.

The simple ascending schema of mineral, plant, animal and human, corresponding with matter, life, consciousness and self-awareness, is common to many cultures and many times, with a general acceptance that the chain might extend to a being or beings above the human, i.e. God. Whatever its limitations as an explanation of reality, such a word picture does offer a better 'feel' for some aspects of experience than the belief that taking something to bits is the best way to understand its purpose. That is because the traditional scientific method cannot tolerate different levels in the nature of reality without the whole reductive edifice beginning to crumble. As a consequence, we have come to see the world from the point of view of the lowest common denominator – that which *can* be analysed and understood in a particular and limited way. We are no longer excited by the mystery of life or the paradoxes of consciousness. It is as if we had become absorbed by the beauty of a rock pool and had lost all sense of the majesty of the sea behind. 'A person … entirely fixed in the philosophy of materialistic scientism, denying the reality of the "invisibles" and confining his attention solely to what can be counted, measured and weighed, lives in a very poor world, so poor that he will experience it as a meaningless wasteland unfit for human habitation.'[7]

That world has come to pass and we do not recognise it. Without a belief in objectivity and in the promise that it can reveal the whole of creation, the 'rational' institutions that now dehumanise people could not have emerged. And without an emphasis on counting and measuring, the money economy – and its peculiar moral blindness – could not have carried all before it, building and re-building those very institutions until they have become a reflection of its own distorted logic. Finally, without the money to fuel its researches and the organisations to provide the collaborative frameworks needed to do ever bigger and better experiments, science could not have kept on producing the discoveries that fuel the constant stream of new products that form the basis of our materialistic, consumer culture. It is an 'Unholy Alliance'. The mind map that liberated our forebears from superstition has become our prison. We do not fully recognise it because we have been brought up to think there is no other. We badly need an alternative vision and one that re-introduces an awareness of the spiritual into our daily lives.

3

Understanding the Question

If you are not part of the solution, then it must be that
you are part of the problem itself.

ELDRIDGE CLEAVER

A good place to start the search for an alternative vision is to
recognise and celebrate just how complex reality is. Paradoxically, science itself is showing us the way. Not the mechanistic,
cause-and-effect science already described, prevalent as it still is in
the popular imagination, but the altogether more subtle and diffi-
dent version that is out there on the leading edge of enquiry into
the nature of the universe and matter. Einstein's Theory of Rela-
tivity rocked the foundations of Newtonian physics by proving
that time and space could no longer be considered absolutes.
Particle physics has shown that you cannot be certain about anything – except uncertainty – and quantum mechanics has finally
removed scientists' fundamental article of faith, their objectivity, by
demonstrating that, at the sub-atomic level at least, observer and
observed cannot be separated, the one inevitably affecting the
other. Implicitly there is a recognition that 'in our experiments we
sooner or later encounter ourselves'.[1] We are beginning to study
the interconnectedness of things rather than trying, artificially, to
separate and exclude them. It is a step that our day-to-day thinking,
still dominated by the certainties of traditional science, must take if
we are to escape from the institutional world that is slowly strangling
our humanity.

This 'interconnectedness', the searching for linkages that exist
beneath the surface, is a fundamental property of a system or
family of relationships, each affecting and being affected by the
others. At one level, this new emphasis brings science firmly back

into the mainstream of religious, even mystical, thinking, where the interrelatedness and interdependence of all things is a constant theme; at another level, it is a new departure, an active rather than a passive intervention, pursuing truth through ever more complex mathematical modelling in an attempt to approximate and thereby predict phenomena as apparently unrelated as climatic change and population growth. The *probability* of an outcome in a specified set of circumstances is the prize rather than an iron law that applies in all circumstances, in all places and at all times.

What is emerging from the continual upheaval in our (largely mathematical) understanding of the universe at both macro and micro levels is the essential duality of nature. In scientific terms form and energy are interchangeable. If an object is to exist it has to have form, but energy is required to both create and maintain that form. Equally, energy without form can have no purpose. Although opposites, there is a complementarity between the two because they cannot exist independently of each other and because both tendencies are everywhere apparent in the universe.

It is one of the profound mysteries of creation that, in a universe where the total energy available is believed to be a constant, there is a propensity towards both greater entropy (disorder) and complexity (order). The cosmos is both running down towards final equilibrium and constancy of form *and* continually throwing up structures that are ever more varied and complex (of which we ourselves represent one of the pinnacles). Pluto (inert, atmosphere in chemical equilibrium, lifeless) can exist in the same star system as the Earth (vital, atmosphere in chemical disequilibrium, life-rich), and which way the coin falls appears to be largely determined by chance. Chaos theories show how, from apparently innocuous beginnings, great events can be set in train; storms from the beating of a butterfly's wings.

The fact that there is order in the natural world is one reason for the success of traditional science, but it is only since the acceptance of the dis-order that is also apparent, and the subtle interplay between the two, that scientists have come to realise the limitations of the old methodologies. Phenomena may exist unchanged for millions of years but there is always the possibility of the kind of event that will create structures of an entirely different order. Human life would hardly have been possible without the rich stew of bacteria that still colonise the earth; but a scientist studying such primitive life forms would be hard pressed to deduce the rich diversity of life on earth today from such beginnings.

In other words, the universe is shot through with symmetries that can be revealed by careful studies but discontinuities exist between such structures that make the links one with another extremely tenuous. It becomes easier to think in terms of hierarchies, with each level having its own internal order and structures. 'The members of a hierarchy, like the Roman god Janus, all have two faces looking in opposite directions: the face turned towards the subordinate levels is that of a self-contained whole; the face turned upwards towards the apex, that of a dependent part.'[2]

Looked at in this way, our understanding of the relationship between the three levels of social reality (see Chapter 1) becomes both more straightforward *and* more complex. We have seen that each dimension clearly has an impact on the others but it is now possible to see that each also follows its own patterns and imperatives. The increasingly all-embracing logic of the structural world may be having an overwhelming impact on the levels of social behaviour and values but there is the clear implication that the assault can be resisted by shoring up the other two planes from within. Tackling the institutional world head on is unlikely to have much impact. We must look to the way we relate to one another and the values we hold dear.

Mind matters

The determinism implicit in the world view of a universe unravelling predictably like a celestial clock (whether divinely inspired or not) has been replaced by a much more fluid picture where indeterminacy, change, transformation and chance rule and where the possibility of the unexpected is taken as read. To make sense of this new world requires a sensitivity to where one actually is and where one is trying to get. Understanding the question is usually more important than finding the answer. Unfortunately in our day-to-day thinking we still respond to the world as if it were a piece of Victorian engineering – complex, beautiful, but ultimately nothing more than the sum of its lifeless parts. It is that leap into uncertainty that we must take if we are to begin to rediscover the mystery in creation and, in the process, within ourselves.

The essence of duality – form/energy, order/disorder – is balance, and balance is now seen everywhere in both the natural and the social worlds, where rhythms and cycles maintain a tenuous equilibrium that is always adjusting to changes in the organism's or system's internal and external environments. It is a dynamic, ongoing process, an organic rather than a mechanical metaphor. The word 'balance' implies the occupying of a position determined by the tension of opposing forces, and the notion of a bipolar world – a world of opposites – dates back to the Greeks. It is also deeply embedded in the ancient Chinese philosophy of Taoism – the yin and the yang – which may have prevented China adopting the unipolar materialism that developed in the West and which was so necessary for the growth of the scientific method. It is through the resolution of opposing forces, the finding of a new balance, that growth and development can occur. To deny the need for change is as dangerous as to pursue it for its own sake.

This tendency towards dualism is rooted deep in our very nature. Our brain – the organ through which we ultimately construct reality – has a left/right asymmetry, each side having its own distinctive perception. Taken together, these two opposing images appear to produce a 'better fit' in relation to the external world because, in evolutionary terms, bilateral structures have triumphed at all levels of creation. The functions of the two hemispheres can be expressed as follows (although this schema, in itself, gives an inherent bias towards the left brain's contribution. Moreover, recent research suggests that the inter-connectedness of the two 'halves' is altogether more complex):[3]

Left	*Right*
logical, rational	intuitive, emotional
verbal	non-verbal
analytical and convergent	syncretic and divergent
sequential	simultaneous
linear	spatial

In certain circumstances and activities it is appropriate for one of the hemispheres to take precedence but life, in general, depends on the balanced perspective that comes from the harmonious co-operation between the two.

One of the most startling features of Western society is the emphasis that has come to be put on left brain functions. In a very real sense the triumph of the traditional scientific method has been the triumph of the left brain over its more ethereal partner. Put another way, it also reflects the dominance of the 'masculine' view-point over the 'feminine' that is only now being challenged. The go-getting virtues that underpin much of Western expansion in the centuries since the Renaissance can be seen as the attempt of the male of the species to push his partial vision of reality to its logical, linear extreme. 'Capitalism, including state capitalism, has been made by man in the image of men, a world of smoke stacks and pylons, pumps and pistons, cannons and missiles; a world ruled by urgencies and aggression, appropriation and conspicuous display. Its most recent associates, militarism, bureaucracy and technology are likewise governed by rigidly impersonal hierarchies and cold cerebration.'[4]

The problem is that pure reason, without the insights provided by intuition, can take on a life of its own, leading step by inexorable step towards a preordained destination even when the evidence (always assuming it can be seen or heard) suggests it would be nonsensical or even foolhardy to go there – which is the way institutions, bolstered by a rational, materialistic world view, think and work. This emphasis on the rational may have been a positive response to dogmatic superstition but, in abandoning our inner eye, we have thrown the baby out with the bath water. Certain truths defy analysis. Indeed, the closer one attempts to pin them down the greater is the tendency for them to hide themselves so that, to the literally minded, they cease to exist.

There is the language of ideas (left brain – linear, logical, convergent, masculine) and what might be called the language of living (right brain – spatial, intuitive, divergent, feminine). It is in the tension between the two (the one unyielding and unchanging, the other mercurial and ever-changing) that the endless process of giving meaning to life is to be found. The attempt to shape the language of living into the language of ideas and then to impose that language on life is perpetual. From the Ten Commandments to the American Constitution, humanity has attempted to pin down the insights gained through life into comprehensive, all-embracing axioms by which to explain, confine and judge what is happening.

Language is, thus, central to the process of change because it both imprisons and liberates. It ties us down to established patterns of thinking and behaving but it can also break through those same shackles by generating and then disseminating new insights. 'Beyond any specific intention which poetry may have…, there is always the communication of some new experience, or some fresh understanding of the familiar, or the expression of something we have experienced but have no words for, which enlarges our consciousness or refines our sensibility.'[5] That is a common enough happening in everyday life for us to recognise the importance of trying to put something 'into words'. The juxtaposition of two familiar words or phrases in an unfamiliar context can produce new meaning which, in turn, can release the energy that always accompanies the creative act. In a very real sense the world is a different place as a consequence. Major change comes when one mind map – one way of viewing the world – is replaced by another. Such times are inevitably confusing, with individuals pulled back by the familiar, yet drawn on by half-formed visions of a better future. We appear to be living through such a period.

Because living is essentially a right brain activity – an endless and intuitive squaring of the circle – any (partial) 'truth' contained in any mind map begins to fray and fall apart as soon as it is uttered (with the consequence that they are being continually amended, updated and ultimately replaced). Through language we daily explore the contradictions that exist between what we believe we are doing (values) and what we actually do. It is a creative attempt to integrate experience and, in the process, address the key questions of existence. Dealing with the contradictions inherent in the 'masculine' and 'feminine' world views, for example, provides an implicit revision of the meaning of 'self', while discussions of 'right' and 'wrong' contribute to the development of morality.

Such debate is vital and requires a vigorous and flexible language that will allow the confusion of preconceptual feeling and thought to see the light of day by giving them meaning. Language is, ultimately, the test of whether ideas are living or dead, whether we are holding on to them for fear of the alternatives or whether they are pulling us forward almost despite ourselves.

The less precise a language, the more difficult it becomes to probe beneath the surface and glimpse the deeper meanings of existence. A diet of buzz words and hype soon ensures that communication becomes stuck in a quagmire of endlessly changing images, removing the ability to distinguish between the nature and quality of the contents. Everything is reduced to the same level, which is the lowest common denominator of mass taste: in other words, what everyone can sign up to and still believe they are making a personal choice and contribution.

By contrast, true communication – the mutual exploring of ideas and experiences that can lead to insights that neither party had been conscious of before – is notoriously difficult, and no more so than when dealing with our perceptions of feeling. It has been suggested[6] that, in the counselling relationship, the 'art' of the therapist lies in being able to engage with the client to bring into being a unique and mutually fashioned language that will allow both to understand the issues that are being presented. Such a language will be rich in metaphor and symbolism, carrying with it a heavy emotional charge. There is a lot at stake, nothing less than the potential for progress towards wholeness or a collapse into a further cycle of failure. It is a subjective and singular journey of significance only to those involved. To convey what has happened in a given therapy is therefore virtually impossible. Yet the attempt to draw universal truths from the particular is vital to the development of the relationships we call therapeutic – and, by implication, to our everyday relationships as well. A language has to be forged and continually adapted and re-formed as perception deepens. It must seek to clarify, include, and encourage a robust and open debate – all properties that are strangled in the rise of a jargon-filled lexicon that both obscures (because it permits laziness) and excludes all but the initiated from participation.

Colour, music, theatre and poetry are yet other examples of the difficulty of giving voice to experiences that are not of themselves rational or material. But, together with feelings, they offer insights into the nature of being and provide a gateway through which to explore the spiritual side of our natures. Given the overwhelming importance of the material in our culture – and the relative ease with which it can be tied down in words – it is perhaps no surprise that the spiritual is withering on the vine for lack of a fertile soil. Most people would readily agree on how to describe a car: the nature of its colour might be a different matter, especially when trying to compare it with other, similar shades. When one moves to more complex images and experiences language can often appear to let us down completely.

Colour is a relatively simple example of a highly subjective ex- perience and one that shows some of the difficulties that occur when we try to communicate a personal response on a general, linguistic level. Fashion is big business, with the major chains insisting on quality, consistency and price. To deliver that degree of uniformity through the design, manufacturing and sales processes requires sophisticated monitoring techniques, which has meant evolving measures for colour matching. When the primary colours green and blue are mixed a unique green-blue shade is created. Material swatches can be compared to an original by saying their hue is too green or too blue, depending on the shift towards either of the constituent colours. Value describes how 'full' or how 'light' the swatch appears against the master, with the colour fading away to white at one extreme and black at the other. Lastly, chroma gives an indication of 'brightness'. Despite these measures, it is recognised that certain people have a better eye for colour than others and, while training can help, there is no substitute for this inherent quality.

And that is true of all subjective experience. Just as we have varying physical skills and attributes, so each of us has a different capacity to understand and appreciate the matters of the heart. In a world where measurement and demonstration are the accepted tests of reality, it is sometimes difficult to accept another's per- ception without such validation. Or else we abandon rationality altogether and act as if anything were possible. That is a failing of language and the thought processes they imply. Without a vigorous vocabulary and a way of making judgements about the world, we can only drift rudderless, one moment highly sceptical, the next uncritically accepting.

The more refined the language, the more likely we are to be consistent and to understand why it is we disagree with others. While we may have little problem acknowledging that some people have a better ear for music than others and can experience emotional and aesthetic depths quite inaccessible to many, we appear today to have more difficulty with religious experience. That is because the language we use to discuss our appreciation of music is well established and reinforced by the evidence of enjoyment displayed at concerts. By contrast, the falling away in church attendance, our consequent distrust of religious convention, and the leaching out of religious teaching from our culture's everyday stock of wisdom, has stripped us of any confidence that we can tell saint from charlatan.

The collapse of established religion hasn't extinguished the impulse for spiritual enlightenment: quite the reverse. An emphasis on the individual, coupled with the established Church's apparent lack of interest in practising what it preaches, may have led to a turning away from dogma but that has only emphasised a desire for personal, here-and-now experience. Not only does conventional religion – of all persuasions – appear lifeless and lacking in relevance, it is out of step with the desire to *feel*: whatever else the human development business has been about it has been about getting in touch with one's feelings, emphasising the need to experience life to the full. Hence the upsurge in evangelical services and the more esoteric New Age experimentation. It is a sure sign that change is in the air when such beliefs attract either adulation or vilification with little in between. Language has failed to bridge the divide and, in those circumstances, there is no alternative to being for or against. That battle for the mind is part of the wider struggle to redefine the relationship between the material and objective and the spiritual and personal that is everywhere around us.

From what has been said above, it should be clear that progress will come through the crystallisation and externalisation of new meaning through the medium of words. A nascent language *is* stirring, although it is as yet scarcely visible among the rag-tag army that has turned its back on the apparently impregnable ranks of the Unholy Alliance. What is clear, however, is that progress will come not from the increasingly rigid and all-embracing mind map typified by the institutional world but through an exploration and sharing of the meaning of our individual existence. It is equally clear that such journeys will occur only through living relationships where trust and honesty allow a deepening of mutual exchange and understanding. That, in turn, implies the kind of environment that we would normally describe as 'community'.

Sooner or later battle will be joined between the Unholy Alliance and the ideals vying to be its successor. If the life contained within the concept 'community' is to carry the day, it will be because the ideas and ideals have been grounded in a vibrant and dynamic language that will make people want to change the way they are living because it seems to be as natural as breathing. Only then will the redundant structures of the money economy collapse and fall in on themselves.

Finding the balance

A recognition of the basic dualism in the nature of existence is also important because it is through the resolution of opposing forces, the finding of a new balance, that individual liberty lies. Without that bipolarity we would be denied the room to make choices. We would live in a two-dimensional world where, once drawn, nothing could be altered. For most of the time we behave as if we did, indeed, live in such a world, not thinking of the consequences of our actions, driven by the winds of fashion and institutional expedience, first one way and then the other, failing to realise that the bedrock of our own being would anchor us if only we could begin to experience it. The adversarial approach – the 'black' and 'white' world of the institution – exaggerates this tendency, making us mistrust our natural responses to the basic messiness of much of real life, and to look to others to provide an apparent clarity in the ever changing kaleidoscope of being.

'Good' and 'evil', 'destruction' and 'creation', 'strength' and 'weakness', 'sickness' and 'health'; there can be no certainty, only the knowledge that life is about the attempt to resolve what cannot be resolved. There can be no certainty even that 'good' will triumph over 'evil', only that the outcome will depend on the choices that each of us is making every day. And, like the beating of the butterfly's wings, the effect of an isolated action can be out of all proportion to the intention. Taking that step into the unknown *can* be significant.

In that process we must learn a crucial lesson. The pursuit of a single goal or ideal brings an inevitable reaction. Life is a continual and creative attempt to reconcile opposing elements and only in so far as we can say there is balance and harmony can we be sure we are on the right track. The closer we strive to approach one pole of a pair of opposites, the stronger will be the reaction. The more an individual attempts to organise their life, for example, the greater will be the impact of the unforeseen. In different people these polarities assume a different significance and the 'problems' faced by each will vary as a consequence. People and/or situations may 'flip' from one extreme to the other. An apparently insignificant event can lead a sufferer from anorexia nervosa to switch from bingeing to starving (and vice versa) without any intermediate steps. This tendency is one reason why it is so important to work at peace (which isn't the same thing as passivity); a war mentality can appear as if from nowhere and only time and bitter experience will eventually lead to a reversal of the polarity.

The 'art' of living is to navigate a course through the particular constellation of opposites determined by one's personality, upbringing and circumstances. The aim is 'wholeness' gained through the process of self-realisation, a synthesis of our material, emotional and spiritual beings that is unique to us. Note that the objective is not 'goodness' or 'rightness'. Indeed, it can be argued that a society based on 'goodness' has to *create* 'badness' out there as a means of continually affirming its value. At a cultural level, the espousal of any one pole at the expense of its opposite will deny the individual the means of expressing that aspect of themselves. As we have seen, the characteristic becomes split off, denied, externalised and projected on to the other.

All peoples exhibit general characteristics, emphases that may reflect dominant philosophies or power relations between individuals and groups. These characteristics will encourage some behaviours and discourage others. A 'caring' society will attract and develop the caring sides of its members' natures but without a competitive, or selfish, element, that society may well stagnate or disappear altogether in times of stress or change when hard decisions have to be taken. The opposite is also true. A society, as Karl Marx pointed out, that emphasises competitiveness and personal greed may transform the world but, with no outlet for compassion, the result would be (he hoped) revolution.

People sense that *our* culture lacks balance. Institutional forces dominate and, as a consequence, social behaviour has been reduced to the largely superficial interactions of the daily merry-go-round on which we are all bit-part actors. One way of illustrating just how skewed our social fabric has become is to list a series of polar opposites (the opposing ends of each dimension) that stem from the narrative thus far:

maleness	–	femaleness
doing	–	thinking
having	–	becoming
material	–	spiritual
destruction	–	creation
mass	–	individual
dependence	–	independence
image	–	content
knowledge	–	understanding
copying	–	creativity
reason	–	intuition
rigidity	–	spontaneity
avoidance	–	responsibility
competition	–	co-operation

It is important to emphasise that neither opposite has any intrinsic superiority to the other. Personal growth is important to the individual and society but a continued emphasis on development is counter-productive: 'greater even than the mystery of natural growth is the mystery of natural cessation of growth'.[7] There is a balance in all things, and an emphasis on growth at all costs – whether it be personal, economic or material – will tend to emphasise doing rather than thinking, with a consequent rigidity in the use of language and the copying of existing ways of doing things.

It is no coincidence that, while the Sunday supplements are preoccupied with the characteristics displayed on the right of the table, the reality most people experience in practice is more accurately reflected in those on the left; nor that the way the attributes fall accurately reflects the division between the left and right brain. We are hiding from ourselves just how far the pendulum has swung. Worse still, the juggernaut of the Unholy Alliance is tending to magnify the trend. Unless and until we are prepared to wrest back responsibility for our own lives, rather than expecting others to sort out the mess for us, there is little hope of halting the headlong rush. The problem is that, at the level where we as individuals have most impact (i.e. at the level of social behaviour), the dehumanising effects of the institutional world, the moral blindness encouraged by the money economy and the diminution of the human spirit implied in the traditional scientific method have combined to make us impotent.

The way forward is to embrace the complexities implicit in the dualistic nature of reality and use them to begin to shift the balance back towards the attributes on the right-hand side. It is important to emphasise that we should not be seeking to eradicate the other qualities but to re-evaluate their contribution to our attempts to achieve wholeness and harmony. We should rather be seeking to utilise that most human of characteristics (and one which continues to mark us off from computers and artificial intelligence), our ability to make connections between apparently unrelated phenomena. There will be losses in the sense that soldiers find their particular skills less in demand during an era of prolonged peace. But the challenge is to find the richness that is in each of us – the peacemaker as well as the warrior – and, in the process, to discover a new purpose in life. It is a win-win situation offering a goal that everyone can aspire to if only they are prepared to set out on the journey.

The nature of problems

One of the first implications of dualism is that we must reconsider the nature of problem solving. In his book *A Guide for the Perplexed*[8] E. F. Schumacher argues that there are two kinds of problems, convergent and divergent. The former can be solved in the sense that, over time, answers to such problems will tend to converge towards an agreed solution or explanation; cars – the answer to the problem of moving a small number of people around speedily and economically – have come to look much the same. These are essentially the problems of the material world in which the scientific method is pre-eminent in producing agreed answers.

By contrast are what might be called life problems, issues that have traditionally concerned philosophers and moralists, and which tend to throw up divergent, opposing solutions. Although a traffic jam is made up of a number of individual cars, a traffic policy is unlikely to be a convergent problem. Intervention and non-intervention are mutually irreconcilable: ban all cars from city centres or encourage people's freedom to purchase and use cars on the assumption that market forces can resolve the issue. The polarising effect is more marked the more abstract – and, it might be argued, the more important – the concern. Medically routine interventions such as termination of pregnancy, for example, become a debate about the nature of life itself.

Because we are so used to the scientific myth that all problems have solutions – or if they don't they're not worthy of investigation – we constantly clamour for answers. To satisfy that need, decision-makers are always fudging the issues, finding compromises that offend the least number of people; and, in the nature of divergent problems, such measures only redefine the problem, throwing up another set of polarities. In other words, the way we habitually solve problems creates yet more problems. The approach has to be both more subtle and less clear-cut. We have to begin to learn to live with uncertainty.

Firstly, we have to recognise that there can be no 'final' solutions, only 'outcomes of the moment' that are more or less successful. Thus, questions about freedom, democracy, equality, etc. can never be finally resolved, although the quality and nature of the debate is clearly vital to us all. Such issues do 'not yield to ordinary, straight-line logic; [they] demonstrate that life is bigger than logic'.[9] We should not abandon the rational part of our being in our attempts to resolve life's problems but we must recognise that relying on logic alone will not produce an answer that is acceptable to all (and that will, therefore, be 'right'). Reason and intuition are in opposition and there is no way of telling in any particular situation which is the superior sentiment. Both have to be listened to.

The most profound understanding of the human condition comes from just such a creative interplay between our intuitive and reasoning selves. It is metaphorical, dependent on images and ideas that offer insights, in the way that the billiard-ball model of the atom suggests reality without ever coming close to defining it. From a strictly scientific viewpoint, for example, the predictive capacities of Freud's theories remain inconclusive. On balance, the evidence that has been produced would have led to the model being binned years ago were it not for the fact that the 'explan-ations' he offered still provide a powerful tool for working with individuals in distress. Equally, his revelation of the role of the unconscious mind and his description of how personality develops have entered the public domain, and the way we view ourselves will never be the same again.

To our rational selves, the dangers in listening to our inner being are obvious. At best, common sense, superstition and tradition go unchallenged because whatever ability we may have to glimpse ourselves for what we are is far outweighed by our capacity to turn a blind eye to what we'd rather not see. At worst, spurious theories – often presented with a veneer of scientific respectability – can gain a foothold and lead people to being treated as outcasts or objects and, ultimately, to genocide itself. But if we can never know ourselves in the sense of developing an objective, verifiable model of what it means to be human, is it not best to acknowledge the fact? *All* knowledge then becomes suspect and 'truth' is a matter of judgement, of the depth of one's insight and strength of character. The responsibility is once more firmly back with us as individuals.

The more inner strength that is available, the more likely a society is to find appropriate, human solutions to the problems that confront it. Conversely, the less practised a people is at searching for truth the more likely it is that large-scale, impersonal recipes will be imposed. The philosopher Karl Popper urged that, rather than searching for certainty in any branch of knowledge, we should be concerned with detecting and remedying the mistakes caused by our limited and limiting perception as quickly as possible.[10] That is why listening to the doubts as well as the counter-arguments is so important. People's feelings matter. They may not be able to express those sentiments elegantly and coherently but they may, ultimately, be more valuable than all the reasoned and reasonable arguments that have destroyed lives down the ages. There should be no assumptions that cannot be challenged. Only in that way can we be sure that the values underlying the rationalisations we offer for our actions are indeed human and life- enhancing.

The particular make-up of divergent problems means that time will be a key dimension in reaching a satisfactory conclusion. Stakeholders need to be canvassed and space allowed for a proper debate to unfold. In our pressured board rooms and committee rooms such a statement would be greeted with laughter. Where is such time to be found? And yet, it is estimated that 'in most companies, half the employees have no idea where the organisation is going, 75 per cent get no feedback on performance and 100 per cent believe management has a hidden agenda.'[11]. It is tempting to suggest that one of the prime reasons why decision-makers *are* so pressured is that they are having to resolve the problems created by their inability to involve people in the communication process; it may be a case of less haste, more speed.

It is also a case of selecting the most appropriate *method* of decision-making. There are plenty of occasions when the simplest and most effective way of taking a decision is to leave it to one person, usually someone who has been given 'authority' to make that decision: choosing the colour of the office wallpaper or running the tea fund should never be left to a committee if sanity is to be preserved! At the other end of the scale, major decisions during wartime are probably best taken by an individual who is able to act with a sense of destiny and as if they were the embodiment of the people who have elevated them; but war is a high-risk strategy at the best of times and is hardly the most satisfactory model for solving everyday problems.

Most organisations, while remaining authoritarian in practice, do flirt with ways of communicating with the people who work in them. Consultation is the most common way of involving the workforce while leaving final decision-making firmly with the boss; it is still a one-way process. Participation is a more complex interchange and allows much more of the decision-making work to be done by the group as a whole. It should be noted however that, to be effective, the number of people who can be involved begins to get significant. Many thousands of people can be told what to do, many hundreds can be consulted but it becomes diffi-cult, though not impossible, for even a hundred individuals to participate effectively. Participants have to both give and receive high-quality, up-to-date information which both dramatically increases the amount of communication that has to take place and makes the task of co-ordination and the time it takes more demanding. At the end of the day, however, participation empowers people, with a resultant improvement in the quality of decision-making, morale and the general health of the organisation. We are also returning to something akin to the potentialities contained in the word 'democracy'.

Most difficult of all to achieve is a consensual approach to decision-making in which authority and responsibility for making decisions are held solely within the group. M. Scott Peck[12] has been a pioneer in the development of 'community', the environ-ment in which consensual decision-making is most likely to emerge. He characterises consensus as

> a group decision (which some members may not feel is the best decision, but which they can all live with, support and commit themselves not to undermine), arrived at without voting, through a process whereby the issues are fully aired, all members feel they have been adequately heard, in which everyone has equal power and responsibility, and different degrees of influ-ence by virtue of individual stubbornness or charisma are

avoided so that all are satisfied with the process. The process requires the members to be emotionally present and engaged, frank in a loving, mutually respectful manner, sensitive to each other; to be selfless, dispassionate, and capable of emptying themselves, and possessing a paradoxical awareness of the preciousness of both people and time (including knowing when the solution is satisfactory, and that it is time to stop and not reopen the discussion until such time as the group determines a need for revision).[13]

People come first, and that is a necessary precondition if we are to counteract the devastation that is being caused by the Unholy Alliance.

We are now in a position to summarise some of the conditions that are necessary for a human-centred, life-enhancing vision to become a possibility. It will require nothing less than the emergence of new meaning from the experience of everyday living. That new meaning will find expression in a language that is both accessible and inspirational by its obvious clarity. It will set us on a road towards a new balance point; one that emphasises the importance of the individual by celebrating the duality of nature, recognises the need to creatively combine the attributes of our left/right asymmetry, and acknowledges the complexities of dealing appropriately with convergent and divergent problems. It is a formidable agenda but one that, once we begin to realise the complexity of life and how disempowering our current methods of solving problems are, becomes immediately a more friendly and inviting prospect; a glimpse of freedom. In a word, it is about building community. All that is required of us is that we set out on the journey.

4

Freedom, Equality and Love

Alice: Would you tell me please which way I ought to
go from here?
Cheshire Cat: That depends a good deal on where you
want to get.

LEWIS CARROLL

'I think we can see the conflict of attitudes which will decide our future. On the one side I see the people who think they can cope with our threefold crisis by the methods current, only more so: I call them the people of the forward stampede. On the other side, there are the people in search of a new life-style who seek to return to basic truths about man and his world: I call them the homecomers. Let us admit that the people of the forward stampede, like the devil, have all the best tunes.'[1]

The best-loved tunes are those from our childhood and adolescence. They may have marked times of great unhappiness but through the distorting lens of nostalgia they become benign, tugging only gently at the heart strings. So it is with the institutions that mark out the social world. Familiarity drains them of much of their threat, both real and imagined. Yet they are the enemy and, given that the power of the institutional world is so overwhelming, it seems obvious that the starting point for homecomers is there. Let us begin the move towards a new 'balance' by describing what a social structure built on human values might look like.

Structures limit opportunities, however; they close down rather than offer openings. It is in the nature of utopias to be unchanging and the very notion of 'balance' suggests something that is in constant flux. If our human potential is to grow through responding creatively to the kaleidoscope of life, we must develop social arrangements that will allow scope for transformation in any direction while at the same time providing an ongoing social identity and coherence; in short, we must acknowledge the polar opposites of form and energy. Form is everywhere around us in the shape of the institutions. Energy will be needed to break through the log jam and that energy will come at the level of social behaviour, the way people behave towards each other, and the language they use, in practice. Change that, and form will have to adapt or become redundant.

For the same reason, it is equally pointless to talk of global solutions if those solutions enslave people. Global co-operation and organisation will, indeed, be necessary but, as problems at that level are invariably going to be divergent in nature, it will be important that those arrangements are consensual. To achieve that means concentrating again on social interrelationships and the role of language, the level where the individual is pre-eminent and what they do matters. No true and lasting change has occurred without its first expression being in the hearts and minds of ordinary people.

As we have seen, 'individuality' is a complex set of ideas that combine a sense of 'wholeness', achieved through self-realisation, within a process called 'becoming' that depends on a deepening awareness of our links one with another. The nature and quality of interactions will be crucial. Put very simply, the content of human relationships can be gauged by the amount of 'love' they contain.

If love seems a strange or even frightening word in this context it is because its usage has become diminished to the point where it is used solely to describe the intimacy between two people who are 'in love' – and more specifically to describe the physical act of 'making love' around which we have so many hang-ups as a society. Beyond that it seems to imply 'demands' that others might have of us; we prefer to keep ourselves to ourselves. In its roots, the word 'love' has a far wider meaning and is in many ways synonymous with life itself, as in 'life-enhancing'. It is worth reflecting that love in this sense has become almost as taboo a subject as death, and in rejecting these two great poles of being it is as if we have become afraid of being alive.

The attributes of love are laid out in the famous passage of Paul's Epistle to the Church at Corinth: 'love is patient; love is kind and envies no one. Love is never boastful, nor conceited nor rude; never selfish, not quick to take offence. Love keeps no score of wrongs; does not gloat over other men's sins, but delights in the truth. There is nothing that love cannot face; there is no limit to its faith, its hope, and its endurance.'[2] In its concern for the other, love knows, enjoys, accepts, responds, affirms and enhances and, in turn, creates the possibility that these qualities will be reflected back to oneself: it contains the potential for growth in both parties. By contrast, it does not attempt to impose itself, it does not compare, dismiss, diminish, or attempt to possess: all reactions that our social structure, based on the Unholy Alliance, encourages and which are essentially life-stunting and polluting.

The ability to love consistently is a sign of maturity; the maturity by which a parent allows a child to find its own way in life – to become itself – rather than imposing their own hopes and aspirations; that allows a partner to explore their potential independently of the immediate relationship rather than insisting that they should express themselves solely through oneself. Love in this sense is rare enough but, in the context of our wider relationships, it is almost non-existent for want of the soil in which to grow. As we have seen, most interactions are played out through the medium of roles, and roles are about power, the power of one individual to define what another can do whether that be in terms of going through a check-out in a supermarket in a prescribed way or the impact that work can have on an entire lifetime by determining what should be done when and how.

Such power allows one person to ignore the other. In large organisations, spanning nations if not continents, this authority is discharged at a distance. There is no need for the executive to come into face-to-face contact with the workers who are being laid off. All that is needed to justify such actions is that the rules are followed (usually with supreme politeness, which is very different from love, as we shall see). The individual is pushed to the margins and their wholeness and uniqueness disregarded. That is why we must convert our 'loveless' society into a 'loving' one.

Otherwise we will continue to live in an increasingly lopsided world where our own individuality can be dismissed with the stroke of a pen.

An emphasis on the individual and his or her relationships is a recurring theme in Western thought from the Greeks onwards. We may still pay lip service to that heritage – especially around election time when certain phrases are trotted out as part of the nation's mission statement – but the reality is that true individuality has become increasingly marginalised and the situation is getting worse. In that sense, we appear to have taken a wrong turning along the way. Where should we retrace our steps to? A high-water mark in our civilisation's thinking about the place of the individual was the Enlightenment, perhaps best summed up in the phrase 'Liberty, Equality and Fraternity'. The French Revolution may not have achieved much for individuality in practice but, at the very least, it might be helpful to rediscover the meanings contained in such words by applying them to our current predicament.

Liberty implies being free from any form of captivity, imprisonment, slavery or despotic control. Today, it is most likely to be interpreted in terms of pursuing one's own self-interest which, in turn, has come to mean spending what one has on what one wants. And with the exercise of that freedom has come the peculiar moral blindness of the money economy. An alternative view of freedom is suggested in the notion of the *freedom to become*. If free will has any meaning it is in the personal resolution of the conflict inherent in being, which stems from the dualistic nature of existence. Becoming, the process by which one becomes whole, is the acceptance of the uncertainty of life and being open to the need to change. Through increasing self-awareness (through becoming), one is able to tap the strength that is at the core of one's selfhood. One can literally stand up for one's self.

From the study of ecology we are learning that there is strength in diversity. A bramble is not an oak but, alone, each is potentially diminished. In evolutionary terms each may be attempting to increase its share of the sun but the system of which they are both a part is weakened if one is driven to extinction. Each has a part to play in maintaining the ecological balance, and the greater the variety of species the more stable and self-sustaining the environment becomes – and the less prone to devastation by outside change.

Each of us contains a seed of being, and 'society' should be about allowing each seed – whether a bramble or an oak – the opportunity to develop its potential, for the ultimate benefit of all. In the final analysis, the process of becoming cannot be prescribed or laid out as a path to be followed; it is something that has to be lived by each of us. In so far as we are not constrained from pursuing that individual path we can say that we are free, because all the traditional freedoms – of speech, thought, assembly, etc. – are implied in it. In describing it in this way we are both setting a clear goal and liberating ourselves from the aimlessness of being free only to spend.

For that reason, the freedom to become is not the same thing as the freedom to do what one wants. It has already been noted that the more self-aware we become the more important and the deeper our relationships with others become. To make free with someone suggests that their needs are being ignored and, for much of history, that has been the position of the strong in relation to the weak. This is why the notion of all people being created equal is seen as being so central to the democratic process. Equality is the bastion against the excess of freedom and vice versa.

But equal in what? Clearly people are not all equal in terms of their natural gifts, their popularity, etc. Taken at a literal level, the Parable of the Talents[3] tells us that if, suddenly, everyone were to start off at the same point in terms of material wealth, it would not take long for inequalities to become evident. And, in attempting to curb those inequalities, how far is individual liberty, as we have just defined it, stifled?

If freedom is the freedom to become, then equality should be the guaranteeing of that freedom to all in equal measure. The bramble should have as much right to establish its potential as the oak, and neither should be seen as something to be rooted up and destroyed. How far is this balance between freedom and equality evident in our society? In terms of access to material wealth it is clear that a doctor will always be better off than a labourer, if only because he or she will earn more for each hour they work. That might not matter if both had equal access to the means to realise themselves; in which the job they did was only one factor on a much wider stage of opportunity. Unfortunately, to be able to play on that wider stage – in terms of leisure, education, participation, etc. – requires money, so the doctor will always have more choices available than the labourer, and the point at which the doctor's ability to make those choices is at the expense of the labourer is always difficult to determine.

It is a sign of the difficulty large-scale society finds in reconciling the fundamental opposition inherent in freedom and equality that the spirit of the law – the means of testing the balancing act in practice – has long since been lost in the institutional interpretation of its letter. The key to this dilemma lies in fraternity. Without any notion of fraternity, freedom and equality become abstractions. One can be free or equal only in relation to others. If those others are merely a mass then freedom and equality become defined on an ever smaller scale, in mechanistic terms such as the freedom to use a public right of way or one's equality in placing a piece of paper in a ballot box.

When the other is known, engaged in the same process of becoming, then freedom and equality take on an entirely different, and in some ways synonymous, meaning. That identification and commonality of purpose makes it as important for oneself as the other that they are free to pursue their goal of becoming; and that very process becomes something that can be discussed and debated so that adjustments on both sides can be made. Both are then truly equal by virtue of sharing that mutual freedom.

For fraternity – or love, to give it a non-sexist, more all-embracing name – to exist there has to be a tie between individuals, a community of interest that binds them together. It is through that identification that the whole is able to become greater than the parts and for the parts themselves to grow as a consequence. It is a virtuous circle, limited only by the commitment of the participants. In the sense that the resolution of the conflicts inherent in polar opposites defines the boundaries of higher-level concepts such as selfhood and morality, so love becomes an ultimate expression of being, an opportunity to explore the mysteries and meanings of life.

Once individuals feel they are not getting their share of those rewards, however, the magic can vanish. Most such ventures perish in that way because they are founded on premises that are unsustainable. In the end equality fragments and freedom becomes divisible. Without love the individual is merely an atom in a hostile universe whose only defence – the right to appeal to an abstract law – is no defence at all.

Stepping stones of the mind

The arguments advanced so far suggest that love can endure only if it is based on the freedom to become, a freedom that is extended equally to all. That, at the end of the day, is a matter of faith, the basis of another, limited mind map, another partial truth. If freedom, equality and love are to be a template on which to build a new model of human interrelationships it is important to ask whether such a society could exist in practice.

To begin to answer that question we need to remind ourselves why we need other people. From the hierarchy we established on our desert island (see p. 7) it is clear that we all have needs. Some of these needs can be met from within our own resources (independence), some we meet with the assistance or co-operation of others (inter-dependence) and some we need others to do for us (dependence). Independence and dependence are a pair of polar opposites and the nature of duality should lead us to conclude that life is about finding a balance between the two. In our culture of rugged individualism, however, independence is the pole stressed, and many people find real difficulties in being dependent in any areas of their lives whatsoever. Put another way, we usually find it much easier to give than to receive, which is one reason why people in receipt of services are disempowered – because there is an assumption that they have nothing useful to give in return that service providers might want.

Paradoxically, despite the emphasis on self-sufficiency, the Unholy Alliance has created an environment in which we are *all* dependent. Our ability to pay for what we need doing disguises that fact but the reality is that we have all become de-skilled in areas where even our poorest forebears were independent. Only when we curse the time it takes to get our washing machine repaired, or when our train is cancelled yet again, do we feel the impotence of not being in control. For the poor, that is a daily experience, and one that is often dismissed as fecklessness, a symptom of life on the dole being too easy. It is obviously more convenient to blame the poor for their situation rather than recognise the implications it has for us all.

During the sixties and seventies there was a vogue for self-sufficiency, a throw-back to the frontier spirit, in which a family took to the elements and survived, reliant only on themselves. The movement took many forms, from 'dropping out' and creating the archetypal homestead to weekend cottages and more bizarre manifestations such as having one's own fully stocked nuclear fall-out shelter. The latter example shows just how illusory and self-ish independence can be, for, even assuming that the individual or family had lived through the initial holocaust and discovery by other starving survivors, any hope of long-term survival would depend on co-operating with others and sharing what resources remained. In fact, apart from basic functions such as feeding, washing and toileting ourselves, *most* of our needs are met most appropriately through varying degrees of interdependence with others. As we move through the hierarchy towards self-realisation the more we need other people in a set of mutual, reciprocating relationships where one person's striving for wholeness is as important as our own. It is in that kind of interaction that true and lasting growth is most likely to occur. It is also the environment in which the challenges of life are most likely to be surmounted successfully.

Interdependence with others occurs in many ways. A dinner party, a wedding or a trial all depend on people interacting. However, once the reason for the individuals coming together ends or disappears so do most of the interactions (unless some individuals are habitually involved on such occasions – for example, the clerks of the court and solicitors – when a clique is formed which outsiders find very difficult to make any impact on). The most productive interactions are those in which people do things together over a period of time. Our ability to enter such relationships depends on our competence in the chosen activity. Competence is one of the main ways in which we express ourselves and through which we gain in self-esteem, whether it be in terms of turning wood into furniture, keeping a house clean or running a youth group. It is a creative, outgoing expression of our selves and, at the end of the day, any society depends on the harnessing of these talents.

We all have different talents and different levels of competence in those skills. They are most obviously used to meet our own needs and those of our immediate family and friends, whether it be in terms of decorating the house, teaching one's children a musical instrument or cooking a special dish. Beyond that we do very little with skills that are not immediately applicable to the limiting set of tasks required by our jobs. Work turns us into professional tennis players with outsize playing arms. Other talents wither for the lack of time and opportunity to develop them. We may have the vague idea that one day we might do more with them but nothing ever seems to come of it. We become lop-sided, partial beings.

So, as well as needs that are generally best met through interdependency, we also have talents. There is a third category of activities that might be termed aversions, needs that we satisfy through our own talents because they have to be done but which we'd gladly leave to some one else given half a chance. For each one of us the combination of needs, talents and aversions will be different but in any reasonably sized collection of individuals there is likely to be a fair degree of overlap; one person's area of need or aversion will be another's area of skill. The challenge is to create the links that would allow the two to come together; to fashion an environment in which needs, aversions and talents could cancel each other out in a mutually supportive way that would also promote individual growth.

One of the main obstacles to such interchange lies within ourselves and our reluctance to lose any vestige of our much-prized independence. We all know people who are good at something we aren't – whether that be in terms of gardening, making curtains, relating to children, mending cars, or the thousand and one other talents that make the world go round – but we're inhibited about asking. Either we don't want to impose because we don't feel we have anything to offer in return or we're afraid of being indebted. By the same token we may be aware of the needs of others, needs that we might be able to satisfy, but we hold back because we don't want to embarrass them, or we fear rejection or over-involvement.

So nothing happens and both we and our immediate community are the poorer for it. At worst our need – or theirs – is left unmet. Alternatively, we pay someone to do the job for us (always assuming we have the wherewithal); we call in an expert with whom our interaction is restricted to the here and now matter in hand. It is a relationship with no past and no future, it is a role: interdependence lasts only so long as we have the means to pay. And note that it is not *true* interdependence, because the good or service is provided as a package; there is little or no room for us to work or participate alongside. It is an attempt to maintain our independence by being totally dependent on someone else.

By comparison, doing something with someone I am going to have ongoing contact with potentially enriches the relationship, even when that something is being done for me, *provided* there is a mutuality about the interaction and it is not one of dependence, i.e. I have nothing to offer in return. It is through building up such a network of mutually supportive relationships that self-realisation becomes a possibility rather than a remote ideal.

In the past, power relations between individuals (i.e. the level of social structure – usually reflected in the possession of wealth) defined who did what for whom rather than any consideration of mutual needs. A peasant worked the land to feed the family and pay an annual tithe to the feudal lord even though, in human terms, the latter had less need than the former: the peasant was dependent on the feudal lord. Today, that exploitation is less obvious but nevertheless ever present. The reason people do things for others is not primarily to develop themselves and their talents, or even an identification with the needs they are employed to meet, but to obtain money to sustain a life-style that is compatible with their self-image. In the process, the meeting of need has become equated with an ability to pay.

Unfortunately, it is usually individuals with the least money who have the most needs in terms of having a roof over their heads, being the least healthy and unable to get away from the impact of poverty. Worse still, the poor are poor because they don't have jobs, not because they don't have talents. They find themselves in an environment where they can't utilise or develop those talents and, as a consequence, they come to perceive themselves as being useless, having nothing to contribute. The bankruptcy of our culture's response to meeting even basic human needs is nowhere more apparent than in its acceptance of this state of affairs as being unfortunate but inevitable.

In other words, there are talents going to waste at all levels of society for the want of the opportunity to use them. At the same time the services that society is able to 'afford' are stretched to breaking point and, because of their institutional nature, often causing as many problems as they resolve. It is a crazy situation that seems to have come about purely because our complex society has found no other way of organising itself other than through the exchange of money. Freedom, equality and love have been sacrificed without our noticing it.

Negotiated settlements
We are constantly in a dialogue with our surroundings, implicitly negotiating our way through life. In every conversation we have we are testing out the limits of action available to us. For much of the time those limits are set by the long-standing nature of the relationships. All family relationships, for example, however unbalanced the power between individuals may be, are the outcome of negotiations, implicit or explicit, and the sometimes violent upsets – when children reach adulthood or during separations – are the result of more or less successful re-negotiations of the status quo. Even when people are hardly speaking to each other there is still some notion of 'give' and 'take', of the need to find a way forward that both parties can agree if not accept.

This inbuilt capacity to negotiate is something we learn very early as we strive to exert our independence (our will to do what we want to do) in a world in which we are, initially, so obviously dependent. It is one of the main lessons of growing up and, as always, the balance is finely poised. People can become either dominant or submissive, and often seek out the opposite in relationships, satisfying each other only in so far as neither can be allowed to grow. That is why human relationships are so difficult and why 'negotiating' has to be a skill we absorb from an early age.

Contrast the subtleties and the potential anguish of the human condition with the simple imperative of the money economy to get the best deal available (and to hell with the consequences for the other) and we begin to grasp something of its attraction. When we can be persuaded that the outward display that comes from our ability to exploit our position in the market place is actually a reflection of our worth as human beings, then it becomes easier to see why we are so ready to avoid entering into negotiations on an even footing. We can avoid taking risks. Money gives us the position and power to ignore the other. The adversarial approach, so typical of institutions, has a similar attraction in that it emphasises issues and not people; *they* can be ignored.

For negotiating to be a successful strategy in human relationships we must all feel satisfied with the outcome. If one individual is always 'winning' at the expense of another, a relationship of power develops and the choices available to both parties are reduced. This is not to say that all transactions should aim to achieve parity. Life is not like that but as long as we feel, on balance, that the deal we are getting is fair and that, over time, we lose some but we also win some, we retain the sense that we are in control of our own lives. It is that strength that allows us to keep going and to cope with the unexpected setback when, for the moment, we appear to have lost everything. Problems start when, for whatever reason, we begin to lose on a regular basis. At first we fight back but, if the situation persists, we will probably cease to notice what is happening. We will accept the reduction in our freedom of action without thinking that control has passed elsewhere. At its most extreme we cease to make any choices whatsoever. That is the situation that many people in total institutions find themselves in.

Love empowers, ensuring that the 'other' in any negotiation is able to achieve what he or she wants as well as oneself. A win-win situation becomes the norm rather than the win-lose that is so characteristic of our competitive world. As the relationships so necessary to satisfying the higher-level needs deepen and extend, so there is an inbuilt and developing awareness of the needs of the other. A virtuous circle has been created that will reach ever higher towards completeness.

Choice, decision-making and responsibility are inextricably linked. We can make the big decisions in life – such as whether to turn off a life support machine that is sustaining a near relative – only if we are continually making choices of lesser moment and accepting responsibility for them. The richer the contract we are negotiating with our environment on a day-to-day basis the greater will be our resources in times of crisis. Conversely, just as our ability to remain in control is a tender flower, so the gradual erosion of choice can go unnoticed. A diet of how to spend our money, whether we ought to move to a different area, or whom we think we can spend a few hours with might fill the day with decisions but they are hardly likely to develop our capacities as individual human beings. What needs doing, how can I make the best of what I've got and whom am I having difficulty getting on with and what can I do about it – these might be more valuable issues to pursue.

One of the problems with the way our society is structured is that different skills are valued disproportionately, and this arbitrary allocation of rewards affects the negotiating position of the skill holders. Our democracy may lay down equal rights for all in certain prescribed areas but, even then, the doctor is always likely to have access to greater resources than the labourer purely because of the relative value placed on their jobs: the housewife, of course, has less power than either. Equality before the law can depend on how much money you have to defend yourself, and your perceived status (which also depends on money): the same applies to access to healthcare, education, etc. These differences in opportunity have nothing to do with the individuals concerned, how good they are at their jobs, how popular or respected they are, or what their actual needs might be; it is dictated by the relative (money) value that is placed on their roles.

Ironically, it is not unknown for doctors to spend their spare time or retirement digging their gardens and getting a great deal of satisfaction from the manual labour involved. Among their friends they may even be known for the quality of their gardens rather than their professional pursuits. Likewise labourers can have the kind of sympathetic personality that 'heals' those in their immediate circle far more effectively than a five-minute consultation with an overworked GP. Yet the fundamental imbalance in their choices or power remains because the skills by which they earn their living are valued differently. If equality of access to self-realisation is to be one of the cornerstones of social interaction based on human values, how can this danger be avoided?

Certain skills will always be prized more than others, and some will have greater scarcity value. Some skills can be mastered with relatively little practice or training; others require a lifetime's commitment to learning. However you look at it life is unfair and, in that sense, equality is a mirage. But the freedom to fulfil the potential within one's self, to become a bramble or an oak, is just that. It does not mean trying to force oneself to become an oak as the only way of being valued, nor to stunt one's development because it is considered undemocratic to be as tall as an oak. Equality comes with the recognition that the bramble and the oak need each other for ecological security and it is in the interests of both to allow the other maximum scope for growth.

To do that means recognising two separate but related problems with the way we currently approach the issue of skills. Firstly, we look at the role not the individual. Just because I am a doctor does not mean I necessarily have the inner qualities for the vocation. I may have been attracted by the pay and status, I may have been pushed into it by my family but, once a doctor, it is very difficult for me to change course. I have to be a full-time doctor or nothing. As a consequence, I may not be a very good doctor but, as long as I do not become a very bad one, I can continue to practise. Most of my patients are either too apathetic to change or don't know a good doctor when they see one, because they never get to know me as a person and thereby begin to understand something of the transaction between us.

By focusing on the individual rather than the role such a mismatch becomes less likely, as any skills I possess will be attached to me rather than the role I am undertaking. People will be looking at the whole 'me', in which my love of gardening and the insights it gives me into the healing process may be a significant consideration. Only within a community where freedom, equality and love are working towards the end of individual self-realisation for all is a differential valuing of skills ultimately meaningful, because it reflects the 'wholeness' of the individuals who make up that community. Choosing between a number of individuals with proven healing skills becomes a process of genuine negotiation, empowering both parties by ensuring that both have commitment to the process and its outcome.

The second problem is to acknowledge that the value given to skills is largely culture-specific. Healers have not always enjoyed the status they do today and, as the legal profession continues to play the negligence card in its assault on the institutional power of medicine, that eminence is probably on the wane. The rewards society divides between its members are almost entirely a reflection of its priorities which are, in turn, linked to the power relations that exist between key players. In an era when making money is everything it is hardly surprising that the publicly successful are those with money, or that the personality traits associated with 'making it' are held up for others to emulate.

In a community within which the equality to become is a central tenet, importance will be attached to ensuring that the basic needs for food, shelter and security are being met. Once these have been achieved and are being maintained, the emphasis might shift to enabling skills. Assisting those who have failed to reach an accommodation through negotiation to find a mutually acceptable way forward, passing on and developing skills in others and helping the next generation to find its way towards its own forms of self-realisation, might be more important than being able to do things oneself. The ability to enter and sustain relationships might be more valued than leadership or the ability to make tough decisions. Equally, the 'gift' of interpreting dreams – to help individuals to be in touch with their unconscious selves – or dreaming dreams on behalf of the community might be considered more useful than a facility with figures and balance sheets. Which is not to say that doing things, making decisions, accounting for expenditure or offering a lead would not be useful skills, rather that they would not have the premium attached to them that they do today. As a consequence, the personal qualities considered important could be very different.

A fundamental problem remains, however; merely switching the rewards from one set of skills to another – from, say, an ability to manipulate the money markets to counselling individuals and groups – does not, in itself, remove the dangers of differential value. It might make it fairer if rewards were allocated to individuals rather than roles (i.e. 'good' counsellors are rewarded rather than all counsellors *per se*), but if the effect is still to give certain individuals an unfair negotiating position *vis-à-vis* others, we are in danger of losing the equality to become. That would certainly happen in the money economy because scarce skills, or ones that required greater training, would inevitably attract higher rates of pay, and the gap between rich and poor would be maintained with its consequent skewing of opportunities for self-realisation.

One way out of this impasse is to recognise that the fact that certain skills are in short supply or require a lengthy apprenticeship does not necessarily mean that they have to be treated differently from more commonplace attributes. The one thing we all share in common is that there are only twenty-four hours in a day. However good a counsellor I may become, I am limited by my waking day. And, if I am to realise myself – as opposed to becoming rich, or fulfilling a public service – I may want to do other things with my time. The same is true of someone who wishes to benefit from my counselling skills. We might negotiate that, in return for an hour's counselling, they would spend five hours digging my garden – in other words, one hour of my skill is worth five of theirs. But, if that was the only 'marketable' skill the garden digger possessed and a similar rate of exchange applied to other services they required, they would be forced to spend all their time doing that one thing with no opportunity to develop in other ways.

Equality in terms of becoming demands the same exchange rate for all skills (i.e. an hour for an hour). It might be argued, however, that such a system would create other problems such as a shortage of supply in key areas. By differentially rewarding certain activities the money economy, in theory, encourages individuals to move into areas of shortfall. This argument assumes that only financial reward motivates people to develop new skills. The low pay in most of the so-called caring professions suggests that this is far from being the case. It also assumes that someone attracted to a particular sort of work solely by the monetary reward is the kind of person who would be wanted to do it.

Given the right environment – a network of mutually support-ive relationships – the freedom to realise oneself, to allow the seed within to develop its full potential, is a powerful stimulant, a life-enhancing urge quite opposite to the sullen face of restrictive practice that says I am not doing anything extra – even if it inter-ests me – unless you pay me more. If skills are in short supply in a community where love is a living reality, that in itself will act as a spur to individuals with an inclination in that area to give it a try.

Far from providing a mechanism through which skill shortages can be overcome, the money economy actually conceals such deficits. We all know that the health service is underfunded in the sense that there is great unmet need. By placing an arbitrary – what we can afford – cap on the health budget it can be demon-strated that, because posts are filled, there are no serious skill shortages and that the system is working efficiently. The irony is that there are many women who have left nursing to raise families who would love to use their skills but can't. Again, within a con-text of a community based upon freedom, equality and love, and an hour-for-an-hour system of exchange, it would be possible to match demand and supply at no extra cost.

Of course, it would be nice if the person who came to me for counselling had a skill that I conveniently needed. We could negotiate an exchange there and then, and that would be the end of the matter. Unfortunately, life is not as straightforward as that. The sick person requiring nursing, for example, is unlikely to be able to offer much in return until they are better, when the nurse's needs might have changed. In an ideal world, however, everyone would pay off their debts somehow. The sick person would recover and repay the individual who had nursed them by spending an equivalent amount of time doing things for them. If they didn't possess the talents that person needed, they might do something for a third party who, in turn, would discharge the obligation to the nurse.

Such a simple and natural society might provide an ideal to aspire to but, in practice, there is a need for a system of credits to keep track of multiple exchanges between several people. I could then use the credit I receive from someone I counsel (which they, in turn, were given for digging a garden for someone else) to negotiate a service from anyone else in our community. Baby-sitting circles traditionally use beans in much the same way. The important point is that they are only of value within the scheme; to anyone else they are only beans. This exclusivity is crucial if we are to retain the sense of community so important to sustaining a living spirit of love.

The money economy is, ultimately, a very complex token economy but it is universal and has become an end in itself. That end, creating profit for those in a position to utilise it, has produced the very problems that we are trying to reverse. As we have seen, money is of use to us only if we have it, and we will only have it if someone else can make a profit out of our having it. Secondly, we have no control over it: the notes and coins I spend today can pop up as part of an arms or drugs deal tomorrow. And finally, it is issued – and therefore controlled – by authorities who are beholden to the power of profit and the institutions that create it.

By contrast, *personal money* – the use of credits within a defined community – is always available because it doesn't exist in the same physical way as *private money*, which has to be guaranteed by arbitrary assets such as a pile of gold. Credits aren't like dollars or yen. You can't pick one up, you can't lose or counterfeit one, you can't hoard them under the bed. In one form or another private money always becomes an end in itself. To avoid this danger the sum total of credits in circulation in a personal money economy is always zero. At any one time there will be individuals in credit *and in commitment*. The stigma attached to being in debt in the money economy just doesn't exist in the personal money economy. It is actually a public service to be in commitment (notionally in debt) because I am increasing the trading activity within my community whenever I get someone to do something for me. No interest can be earned or a profit secured, so there is no incentive to hold onto personal money. The quicker it circulates, the 'richer' the community is because more goods and services have been exchanged.

Personal money is issued in, and stays in, the community that creates it, so there is a greater measure of involvement in how it is used. The ethical use of personal money becomes a daily possibility. I don't have to do something I don't feel happy about – such as working in a nuclear power station – just because there are no alternative sources of employment. Everyone is free to do as much or as little as they want, when they want, for whom they want. No one need feel put upon, no one need feel indebted. And personal money is created *when it is needed*, rather than having to wait to have a need met until one has the money to pay for it. Equally important, it is an open system in that everyone has access to the trading history of other members. Trust is fundamental to a community based on freedom, equality and love, and seeing what people are doing in practice promotes honest discussion as well as minimising the possibilities of abuse.

Such token economies already exist in embryonic form and have been shown to work in practice over a number of years. Schemes have been set up throughout Europe, North America and Australia along the lines of the Canadian LETSystem – Local Employment and Trading System – pioneered in British Columbia by Michael Linton. Declining demand for forestry products – on which the region relied – had crippled many local communities; 'when recession strikes, so much money disappears from the scene that even very local transactions like selling firewood or hiring baby-sitters come to a grinding halt'.[4] The creation of 'green dollars' allowed transactions to continue and even diversify – from plumbing to flower arranging, from language lessons to clothing manufacture. The experience of a local dentist is typical: 'when patients with hard cash are scarce, [he] accepts, at least in part, green dollars for his services. As a result, a lot of people who can't afford good dental care are now getting it. In return [he] has furnished his office with locally made furniture'.[5] Because everyone can do something, it means matching needs and skills without the question of affordability entering the equation. There need be no such thing as poverty in a personal money system.

Apart from a book-keeper and some means for matching members' skills and needs – both of which can be 'paid for' within the system itself – the organisation is simple and self-sufficient. There is no need for a hierarchy, a group of individuals making the decisions, or a complex politico-judicial system to ensure that individuals' rights are not trampled on. There is no bureaucracy that begins to take on a life and meaning of its own, only a group of individuals taking responsibility for the kind of community they want to create.

It sounds almost too good to be true. At one stroke we have apparently eliminated our dependence on the money economy and have reduced the need for a social structure to a network of human proportions. Freedom, equality and love are in with a chance. There are obvious problems, however. It is essentially a system for exchanging skills or labour between individuals. As such it is geared towards what we currently call 'services' and, although repairs and maintenance fall quite easily under this heading, there is a difficulty in how to 'pay' for any materials or energy consumed, as well as how to get these resources from where they are to be found – or created – to where they are to be used. For the same reason manufacturing is almost entirely outside its remit.

Another obstacle to the implementation of a simple exchange model is the question of communal activity and responsibility. If a system of exchange based on human relationships was to become the basis of community life, most of the duties currently performed by government – both central and local – would have to be transferred. The building and maintenance of roads, public transport, the disposal of rubbish, the provision of libraries, recreational facilities, etc., etc. could not be handled through individual negotiations: they are communal in nature.

While setting up LETS-type trading networks becomes a necessary first step on the road towards creating a network of locally based relationships that offer the possibility of self-realisation for all, it is clearly not enough. There will be a need for an overarching community structure to take responsibility for activities that fall outside the scope of individual exchange. We are moving into the realm of social structures and its inevitable counterpart, the differentiation and separation of power and the consequent allocation of roles. How such issues are resolved can clearly influence how far individuals are able to negotiate on the basis of equality (i.e. how much they are in control of their lives). If certain positions have great power, those individuals could be placed above the need to negotiate on an even footing.

We are beginning to construct institutions and, just as the American Constitution consciously avoided centralising power through the mechanism of representation, so safeguards will have to be written into any community bill of rights to ensure that no institution can come to dominate the agenda. The balance between form and energy is critical.

5

Building a Sufficient Community

Nothing in life is worth doing that can be achieved in
a lifetime; therefore we must be saved by hope.
Nothing which is true or beautiful or good makes
complete sense in any context of history: therefore
we must be saved by faith.
Nothing we do, however virtuous, can be accom-
plished alone: therefore we must be saved by love.

REINHOLD NIEBUHR

Thus far, we have concentrated on relationships between individuals – the level of social behaviour – and the kind of
exchanges needed to encourage a community based on freedom,
equality and love. This emphasis favours energy at the expense of
form which, given the need for balance in all things, is likely to be
as dangerous to healthy development as is today's preoccupation
with the ever-increasing scale of social structure (form). Borrow-
ing another scientific analogy, it is also likely that, in achieving
balance, structures will evolve that are ever more varied and complex – it is another sure sign that present trends run against the
grain when institutions are merely getting bigger and more similar
rather than increasingly intricate and diverse. To counteract that
trend it will be necessary to continually review the values we are
attempting to promote by checking them against what we are
doing in practice. It is a feedback loop that offers the possibility of
fine tuning both our individual and our social behaviour. If our
actions are out of step with our values we have two options, either
to change the way we behave or to look again at our values.

In trying to imagine a social structure based on freedom, love
and equality, one of the first questions that needs to be addressed is

'how big should it be?' The larger the grouping the greater the power that can be concentrated in a few hands and the less identity (love) there will be between individuals. Too small, on the other hand, and there are not enough people to be communally viable. Once again, there is a balance to be achieved and there is little enough information or experience on which to make a judgement.

The independent Greek *polis*, or city state, was small by today's standards of local, let alone national, government, with a population measured in thousands, of whom only adult males were considered citizens. Aristotle thought that the size of communities should be determined by the number of citizens who could be addressed by a single herald – so communication is important; the possibilities of adapting existing and potential technologies to community are obviously much greater today than they were two and a half thousand years ago.

Even Athens at the height of its power had a population of only 250,000 (including those in the surrounding countryside) of whom perhaps 40,000 had the right to participate in the affairs of state. But, although the expectation that every citizen would contribute released energy and commitment, it also led to conflict, intrigue and factional in-fighting. Both size and the lack of an underlying and unifying philosophy (value system) that put the individual centre stage contributed to this falling away and to the eventual creation of a more traditional state.

Erich Fromm, in *The Sane Society*,[1] proposed that face-to-face groups (numbering about five hundred members each) should become involved in the decision-making processes of the (potentially global) state. Each group would be provided with the relevant information relating to a particular issue and, after a specified period of deliberation and discussion, would vote. The aggregate votes would provide a basis for decision-making on that issue. Fromm believed that such a system was inherently superior to the conducting of opinion polls because it brought individuals and ideas into contact with each other; it was an active rather than a passive gathering of opinion.

It remains a top-down system of involving people, however, if only because the issues to be debated, the way they are framed and the interpretation of the 'results' would have to be done centrally. In the work he has done on building community, M. Scott Peck[2] has successfully 'led' groups of three to four hundred people 'into community' – that is to say, the state where the individual and community become one and decision-making is genuinely consensual (there is no 'majority' or 'minority'). We will return to the issue of building community but, for the moment, it is sufficient to note that quite large groups can operate without the paraphernalia of political parties and voting, with all the opportunities for conflict – dividing people – that such potentially institutional and adversarial mechanisms can engender. Building community may be a time-consuming and energy-consuming process, inappropriate for much day-to-day living but, as a means of creating and sustaining a sense of individual participation and responsibility, for moving forward as a community, it offers unlimited potential; we are truly moving into uncharted waters.

At the end of the day, sheer numbers are not the critical factor. Although we are clearly talking about small groupings, there is room for experiment and communal preference. 'Successful' communities may attract outsiders, which could create tensions as the number of individuals grows. Ultimately, the need to maintain a human scale is paramount; it is probable that expanding communities would sub-divide like cells to form two similar entities, while less flourishing groups might wither away and die.

This emphasis on scale raises an immediate question about the role of technology in a community setting. We have grown used to processes becoming ever more expensive and complex. It took ten years and some seven hundred people just to design the pressurised water reactor at Sizewell B.[3] The wealth and welfare of whole towns[4] are tied to a single product, and only the most affluent nations can afford the concentration of resources needed to put people and equipment into space. How can this trend be squared with the ideal of maintaining a scale that is human?

The issue of *how* things are produced will be dealt with in more detail in the next chapter but it is important to stress two aspects of production that effectively turn current priorities on their head. Firstly, while a sufficient community can never turn its back on technology, it will be interested only in processes that are life-enhancing rather than life-threatening in the broadest sense. The on-going development of hi-tech products (including satellites) that facilitate communication will be just as important as at present but the overriding consideration will be the way in which things are produced rather than what is produced.

Secondly, the meeting of basic needs through personal money will ensure that people do not have to undertake tasks merely to afford the necessities of life. In other words, the current pre-occupation with a job at any price – including a working environment that damages individual health or causes widespread pollution – will be replaced by a desire to do something because it utilises one's skills and leads towards self-realisation as well as contributing something to the community.

In certain instances, because of the availability of raw materials or because of the nature of the production process, it is possible to envisage a community based around certain products. Individuals would be attracted because of the specific skills needed (e.g. the scientific orientation of a space programme). The need for the community to be sufficient (a mutual exchange of skills, needs and aversions negotiated on the basis of an hour for an hour, within the context of participation in, and responsibility for, the overall direction of the enterprise) would ensure that structures didn't emerge – such as management and unions – that tended to be institutional in nature, denying the centrality of the individual. Such a community doesn't imply that there is no need for a division of roles that may carry differing responsibilities on a day-to-day basis – clearly, there are times when a doctor is assuming much greater responsibility than a labourer – but that these structural necessities don't lead to individuals or groups being marginalised by virtue of the roles they choose to undertake.

The size of a community is likely to affect the ability of every member to negotiate with their environment on the basis of equality. Where love is present, relationships will be notable for their warmth and sensitivity to the needs of the other but, as noted earlier, it has been the universal experience of humankind that individual interests conflict and compete. In our institutionally

oriented society, what is essentially an individual, human transaction is caught up in the theatre of social and political
stratification where issues of power and position are more important than considerations of equity and truth. For a sufficient
community to grow requires nothing less than that every member
believes themself to be important while, at the same time,
recognising the value of every other member; equity and truth are
fundamental issues that must be resolved on a day-by-day basis if
progress is to be maintained.

To retain the human element, every member of the community
should have the right to involve an independent, third party to
mediate in the event of a dispute. Such individuals would probably
emerge spontaneously in a community but they might also be
formally elected for specified terms. As well as an ability to listen
to both sides, such intermediaries would have to speak with
authority, i.e. be listened to. That authority would come from their
personal integrity rather than from the occupation of a position.
The more individuals with these qualities that a community can
call upon, the stronger and healthier it is likely to be.

To be able to discharge such an exacting responsibility implies
someone who has so thoroughly internalised the values implicit in
freedom, equality and love that they are literally living them in
their daily lives and are therefore able to respond instinctively to
any experience presented to them. The traditional Jewish rabbi
undertook something of this duty, being routinely asked to inter-
vene in disputes concerning matrimony, property and all the other
vicissitudes of life. In the rabbi's case, the frame of reference was
the Law of Moses, its implicit value base provided the guiding
light – although in most instances simple human compromise
invariably won the day. The wisdom needed for such mediation
requires a depth of experience associated with maturity, which
becomes a virtue to be prized in young and old alike.

Most people are able to represent themselves in the day-to-day
negotiations of life and to decide when they should appeal to a
third party. But there are those who, by virtue of ill health, disability
or other factors, are unable to negotiate on an equal footing either
temporarily or permanently. People in this position often require

assistance to do the simplest tasks, and the person offering the support can use their position to dictate terms; it becomes a relationship of power. To restore equity, individuals should have access to an advocate, someone who will act on their behalf in any negotiations concerning their well-being. A tripartite relationship would develop between supported, advocate and supporter, and the interests of all three would have to be acknowledged in the negotiations to determine how the needs of the supported can best be met.

As the ability to communicate needs and feelings – the main ways in which the outsider perceives whether needs are being met – diminishes (as with a debilitating illness or severe learning disability), so empathy and commitment on the part of the supporter become crucial in responding sensitively to the needs of the supported. In such circumstances it is also possible for the supporter to become exploited (either consciously or unconsciously) and the ongoing negotiations with the advocate acquire an added significance in determining what is in the best interests of everyone involved. All parties need to be as honest as they can about their motivation and, through the trust that such opening up brings, to develop a relationship that is mutually supportive without ever losing sight of the conflicts of interest that exist.

Life is never straightforward and interests are often opposed. A previously loving couple may be unable to respond to the change in the relationship brought about by one partner's sudden and unexpected disability (through, for example, a stroke). How does a community resolve a situation in which the able partner is saying they cannot cope and that the disabled partner should move elsewhere? Relationships in today's world are so fragmented that it is unlikely that the couple will have the network of support that might give them the space to decide what it is they, separately and jointly, really want, so a demand for action may, in fact, be a cry for help. In all likelihood the situation would deteriorate until one or more agencies became involved. Social workers might talk to both parties but would inevitably be seen by both as being there for the partner with the disability; their ultimate response might well be dictated by the resources available rather than an objective assessment of the needs of either individual. Lawyers would see both parties separately and the adversarial nature of proceedings would tend towards a polarising of positions. Counselling might reveal the possibility of reconciliation but would be powerless to summon up the support needed. Different 'experts' – all previously unknown to the couple and, therefore, untried – working to different agendas and time-scales will tend to confuse and overload the situation, and a break-up that neither party may really want will become an inevitability, leaving a residue of unresolved guilt and anger.

A community setting provides a much more flexible and co-ordinated response to crisis. For a start, relationships developed through the use of personal money will be qualitatively different – recognising the importance of self-realisation for the other as part of everyday transactions. The process of negotiating on the basis of equity would give the parties confidence that ways will be found to resolve difficulties and might lead them to talk to each other in a different – more open and assertive – way. If either partner required assistance (time away from the other partner, someone to talk to about the changes that had taken place, etc.) they could use personal money immediately without having to worry about whether they could 'afford' it or whether they were 'imposing' on their immediate network (because they would still be negotiating on the basis of equity). If a doctor or other practitioner was involved they could dispense 'credits' to either party or both, and involve an intermediary or advocate as appropriate. The emphasis would be on empowering the individuals concerned, giving them the means to adapt to their changed circumstances in a way that, whether they ultimately chose to remain together or not, leaves them both in control of their lives and able to deal with their relationship in a human way.

This example illustrates the limitations of a simple token economy. There are times in all our lives when we need the support of the wider community; when, in fact, our ability to control our own destiny has been knocked awry. In this instance, the 'doctor' has dispensed 'credits' and arranged for specialised assistance to help the individuals concerned find their feet. Where do these 'credits' or resources come from and how are they ultimately controlled? Even if a community-run insurance scheme operated there would still be the need to administer and monitor how it was used. Decisions would have to be made on the level of contributions needed to balance anticipated demand. The matter of who could draw on the programme, and to what limit, would have to be continually monitored to ensure long-term viability.

Although everyone, no matter how disabled they might be, has potential skills to offer in an exchange system based on personal money, there are some individuals whose needs will always outweigh their capacity to contribute. An advocate may conclude that a level of support is needed that the individual cannot hope to meet through the normal workings of a token economy. How should such support be delivered and who should be responsible for making such decisions? To answer these questions we need to look at how a community might structure activities that cannot be undertaken through a system of individual exchange.

Rights that determine responsibilities

In the West we have come to take rights for granted and to forget that duties are a natural consequence. We pay our taxes and believe we have done our duty; it becomes someone else's responsibility to use those finances wisely. When things go wrong – a child killed by abuse, the collapse of a building, etc., etc. – we instinctively look for someone to blame. Often reluctantly and irregularly, we turn out for local and national elections to register our own limited self-interest. And that is the end of the matter unless we happen to find ourselves in dispute with some bureaucracy, when we might enlist our MP or councillor to defend our interests. It is essentially a private, self-centred life with few obligations beyond those to family and friends. Wider society is kept at bay by our ability to buy it off.

If, however, we look at what that wider society (as opposed to the state) provides, the nature of our privilege – and our lack of social responsibility in recognising it – becomes clear. Our lifestyles could hardly be sustained if it were not for the range of goods and services available. Quite apart from the food that appears miraculously and in such abundance on our supermarket shelves, there is the daily injection of gas and electricity that keep us warm and power a range of appliances from fridges to televisions. There is the telephone to keep us in touch with one another and the plethora of other services, from the milk delivery to the refuse collection, that we take for granted. All we have to do is pay and, as we have seen, that means accepting no responsibility for anything that might have happened to enable us to enjoy what we have paid for. Few potentates have been quite so heedless of the conditions under which their slaves were forced to work – or if they were, they have been quite rightly labelled tyrants by history.

In a society where concern for the other's right to self-realisation is fundamental, such short-sightedness would be unacceptable. The only way of ensuring that what is done in our name – and running inherently dangerous or polluting power stations to provide electricity, for example, is being done in our name – is, ultimately, by our being involved in the process and having ownership of it. This sense of having a stake in what is going on can be encouraged in two ways. Firstly, by involving everyone, however indirectly, in the processes that sustain the meeting of our basic needs – for food, warmth, shelter and security – and, secondly, through the decision-making processes.

To be involved in a process means having direct, first-hand experience of what is happening. The more sufficient a community becomes in terms of basic needs, the more likely it is that members will be aware of what has happened and who has been involved in satisfying a particular need. In deciding whether to accept an egg produced within the community, I am more likely to know the conditions in which the hens were kept. Not only that, I may be able to weigh up such apparently extraneous considerations as the attitudes or skills of different hen-keepers to their birds.

A commitment to self-realisation in myself and others means that a whole range of issues become potentially important. In other words, the transaction is about more than a here-and-now exchange and, given such an underlying approach, I would be unlikely to tolerate an egg produced in a factory farm – which is a consequence of money transactions being about getting the best price or cheapest egg – regardless of the consequences (and for most of the time, we are ignorant of the consequences and care less).

Not all transactions will be as straightforward. Several people may be involved in a process over a period of time. Rubbish collection, for example, is currently a multi-layered activity. Each individual household is responsible for assembling its refuse to be collected by the council and, after a number of operations, it is eventually dumped at a site far removed from its point of origin. Our individual responsibility begins and ends with that initial transfer and, assuming the dustbins are emptied cleanly and efficiently, we give little thought to what we dispose of, where and how it is dumped and in what quantities; in short, we are not involved.

An alternative approach would be for a group of individuals – from the same street, for example – to negotiate with one or more of its members (allocating a number of credits according to the time involved) to take the group's rubbish to a central point *within* the community for disposal in a mutually agreed fashion. Such a simple step is likely to radically alter our view of rubbish, making us look at what we purchase in the first place, what is repairable and recyclable, and what will need to be disposed of and how. Individuals who are acting directly on behalf of others known to them would also be more likely to take responsibility for what they were doing (not just seeing it as a means to an end – getting money) and to be accountable. And there would be no need for a separate organisation employing full-time workers, etc.

If the group of individuals became dissatisfied they could re-negotiate or find alternatives (each taking it in turn, for example). If rubbish disposal in that street became a health issue because no one would accept responsibility the community might, ultimately, have to step in, imposing 'fines' and offering the credits produced to others to take over the task. Such an intervention should be the very last resort, however, because it would be an acknowledgement that the individuals concerned had abnegated their responsibility as citizens. It would be, potentially, the first step on the slippery slope that leads back to an institutionalised world and a divided self.

For all members of the community to be involved in the decision-making process implies committing time – which the community might reward through credits. How that time is spent, whether in large-scale community building exercises, using the opinion-forming groups suggested by Erich Fromm, disseminating information and canvassing ideas through communications technology, or a combination of all three, is up to the community itself – different circumstances will suggest different solutions. If self-realisation is about accepting responsibility for both the self and other, however, participation must be meaningful. The individual must have the sense that his or her viewpoint is valued and significant, and that can only happen if the decision-making processes are open and flexible. There will be occasions when it is appropriate for the community to go with a decision that has been delegated to an individual to make. At other times *only* the consensual route will ensure that a community is moving in the right direction. In between, a range of participatory and consultative mechanisms can be employed. The trick is working out which is most likely to work in a given situation. What is clear, however, is that certainty is a dangerous thing: 'every good thing stands at the razor edge of danger and must be fought for at every moment'.[5] Mistakes need to be detected and remedied as soon as possible.

Such mechanisms can handle the executive decisions facing the community – the major 'policy' decisions that determine which track is to be followed – but the day-to-day implementation of those decisions is best left to individuals or small groups. Groups have the advantage that, if open and flexible, they are more likely to make balanced choices (on the other hand, if closed, they are more prone to veering towards the extremes). Small groups can also remain 'in community' with relatively little effort, but recognising where a group is coming from in practice can be difficult. To encourage openness such groups and individuals should face regular 're-selection' (alternatively every member of the community should take a turn at serving on one of these groups). They should also have reference groups monitoring and mirroring their discussions: not vetoing or criticising decisions but feeding back dimensions and alternatives that may have been overlooked.

These 'shadow' groups would provide experience and broaden involvement in the community's affairs as well as cementing links between members. Individuals with proven skills in organisation, negotiation, etc. might be encouraged to serve in particular groups as co-opted members, but it should also be recognised that over-reliance on any one member potentially devalues the contribution of the rest of the community and limits the self-development of others.

A group of people 'in community' recognises the importance of *all* its members. Listening to the uncertainty of individuals, and encouraging them to give voice to their concerns, becomes as important as acknowledging the confidence of others. Balancing rational argument against feeling, the desire to act against the need to think, and the holding on to uncertainty in the face of the pressure to come to a decision are possible only in the context of a community where true communication is taking place. The more honest and open such sharing becomes, the more likely it is that the conflicts inherent in divergent problems will be creatively resolved, producing 'outcomes of the moment' that are healthy responses to life rather than a mere papering over of the cracks.

Scott Peck identifies four stages of community development – pseudocommunity, chaos, emptiness and, finally, community itself.[6] In pseudocommunity, participants 'pretend' to be in community by being polite to one another, thereby avoiding conflict but also the reason for them being together: it is a way of denying the differences that exist by sticking with generalities and cocktail party confidences (which reminds one of the detached face of institutions that can smile even as they condemn). Chaos emerges when someone begins to get beneath the platitudes, not in an attempt to understand where the other person is coming from but to get them to agree with their own point of view (or, alternatively, to take on the role of 'healer' and 'solve' the other's problems for them). Not surprisingly, people take exception and soon you have a room full of people trying to 'convert' and/or 'heal' each other – result, chaos! It is essentially very egocentric behaviour, with each individual attempting to impose their will on the group. As such it is both ineffective and demoralising.

'There are only two ways out of chaos … one is into organisation – but organisation is never community. The only other way is into and through emptiness.'[7] Emptiness is the state in which individuals have emptied themselves of the many barriers that exist to true communication – expectations, prejudices, ideologies, the need to heal and convert, the need to control, etc. It is a giving up of apparent certainties and securities, even of one's whole orientation to the world, that is both painful and frightening. Without that surrender, however, it is not possible to experience the rebirth that is the potential of community.

> Whether sudden or gradual, however, all the groups in my experience have eventually succeeded in completing, accomplishing, this death. They have all made it through emptiness, through the time of sacrifice, into community. This is an extraordinary testament to the human spirit. What it means is that given the right circumstances and knowledge of the rules, on a certain but very real level we human beings are able to die for each other.[8]

And what *is* on the other side? Firstly, a sense of peace. The struggle is over and one is no longer alone. One has come home. There is also a feeling of increased energy and possibility. Emotions run close to the surface so that laughter and joy, sadness and grief flow without inhibition. Above all, people share. They listen to and feel for one another so that, in a very real sense, they become part of one another and, thereby, of something greater than themselves. Love is at everybody's shoulder.

It should be stressed that communities are not conflict-free, but struggle is both constructive and creative. The process is inclusive in the sense that people come first, rather than being a matter of winning for the sake of winning or to maintain one's own self-esteem. A lot of time and energy may also be needed to maintain community but it offers an environment in which it is possible to live in consensus, with everyone contributing to, and having a stake in, the community's agenda.

The phrase that seems to best describe a community that – in terms of its size, decision-making mechanisms, acceptance of responsibility, quality of relationships, involvement and access to mediation – provides enough to meet the basic needs of its members and allows the opportunity for their self-development is a 'Sufficient Community'. Not, you will note, self-sufficient, because for communities, like individuals, interdependence, not independence, is the natural, growth-promoting state.

The nuts and bolts of community

The effective management of the personal money economy is clearly central to the well-being of the community and can serve as an example of how such a system might work in practice. The executive decisions – taken by the whole community, 'in community' – include how much needs to be raised to sustain the community's proposed activities, the principles according to which each individual should contribute, the allocation of and accounting for resources committed to individuals or groups responsible for particular functions, etc. A group or groups – with one or more reference, or 'shadow', groups supporting it – would be constituted to take responsibility for the implementation of the decisions taken and would meet regularly to discharge their function. Technology would facilitate the task by providing information that would be available to any member of the community at the press of a button. 'Open' meetings might also be held to canvass a wider view on specified issues.

'How much do I have to contribute?' and 'what do I get out of it?' are two basic questions that will be of interest to all of us, however far along the road to self-realisation we might be. Unless the two are in some kind of balance there is a danger that the individual will feel exploited – or be perceived by others to be exploiting them. Participation in the decision-making process and access to information about what is happening will obviously go some way to ensuring that people understand and have ownership of the structure they have created, but what does contributing and receiving mean in practice?

If I am asked to contribute hours or credits to the community, I have two options. Either I can offer the credits I have received through the use of my skills and these can then be re-allocated to others to undertake the communal tasks that need to be done; or I can do those tasks myself. In the same way, the tasks that the community wants doing need to be defined and matched with the resources available, in terms both of the credit budget agreed by the community as a whole and the skills that are being offered.

The example of rubbish collection suggests that many activities currently undertaken by 'specialist' departments of statutory agencies could be done more effectively through informal arrange-ments between sub-groupings within the community (based on neigh-bourhoods or streets, for example) but many tasks will still have to be co-ordinated centrally. Maintaining the electricity and gas sup-plies, mending roads and drains, etc. require a wider perspective than that available to the individual or the street.

At present we operate on the assumption that it is someone else's responsibility to do these jobs. Not only do we have little idea of what these activities entail but we also fail to appreciate the costs in terms of health, pollution, impact on family life, etc. that are created. It is as if such tasks were beneath us. They are perhaps seen as being menial and lacking in status. And yet there are voluntary groups, drawing people from all ages and backgrounds, who devote themselves to restoring canals, rebuilding steam loco-motives and trams, creating proper paths at sites of scenic beauty – all heavy manual labour. In many cases people pay to participate in such activity 'holidays'.

The difference is that in one instance we are talking about a job of work which, in the money economy, defines who we are to others; in the second we are talking about interests which define who we are to ourselves. The distinction is crucial because, in a community setting, the emphasis will be on self-realisation rather than earning a living, and labour and the use of our whole bodies will be a significant part of that process. Another major difference between communal activity and work in this context is that we can negotiate how much or how little of an activity we will do; we do not have to sign up for a forty-hour week, fifty-two weeks a year. Looked at from this perspective it is perhaps easier to see how the installation of a neighbourhood heating system, for example, might be viewed as important in itself, mobilising the whole com-munity in one way or another, rather than menial work reflecting a low status activity. *How* such schemes might be financed through the personal money economy would be the responsibility of the group constituted to implement decisions and priorities agreed by the community as a whole.

Apart from managing the personal money economy, other responsibilities that could be dealt with in this group/reference group way might include health and welfare, communications, education, housing and self-development. It is crucial that, whatever groupings are formed to progress these issues, the emphasis should be on achieving a breadth of vision rather than focusing in on a narrow sphere of activity. Thus, health and welfare would cover everything to do with individual and communal well-being and might include consideration of diet, water cleanliness, rubbish disposal, recreation, etc., as well as some of the more traditional areas of health and disability. Such bodies should be seen not as legislating but as *facilitating* change within the community by increasing awareness of issues. Responsibility for action must remain with individuals and the community as a whole.

Examples of such 'interest' groupings exist today in the form of co-operatives, estate management committees, parochial church councils, etc., and one of the abiding complaints is the lack of people coming forward to do their bit: it always falls on the same few committed shoulders. This lack of involvement is a potential pitfall in any community but in today's society it is endemic and stems from two separate but related causes. Firstly, the activities listed are marginal to most people's lives and reflect the fact that no sense of community or common identity exists; each individual is isolated by the fragmentary nature of their relationships and activities. Secondly, real power lies elsewhere, and participating on committees and action groups does not alter the basic powerlessness that people feel; small victories may be achieved but they cannot stem the tide of apathy that modern life instils.

By contrast, a community that has both a sense of its own identity and controls its own destiny in practice, is more likely to evoke a positive response from its members. They will be engaged in, and care about, what happens. At the very least, they will have an interest in those decisions that might affect their well-being – how much they are going to have to contribute and what benefits they might enjoy; at best they will recognise that the 'wealth' of the community depends on the contribution of each and every member – the greater the number of interactions, the greater the potential for self-realisation for everyone. The longer the history of success the greater such commitment is likely to be.

Going back to the idea of life as a process, it may be that at certain ages certain activities are more appealing. Physical effort and sustained communal activity may attract younger people, counselling and advocacy the more advanced in years. What cannot be emphasised enough is that everyone should have the opportunity to participate in the full range of communal endeavour and have that contribution valued. That does not imply flitting from one thing to another or dropping in and out when the mood suits (for example, only doing manual work when the sun is shining!) but should reflect a willingness to become involved, to test one's skills and develop one's whole being. That desire to commit oneself is ultimately the only safeguard against exploitation of self, others and the environment because it provides one, as a free participant, with the means to say 'this is wrong' and to do something about it.

A little understanding – neighbours
Even assuming that our community is now firmly on the road to freedom, equality and love, it can never exist in isolation. Just as individual relationships are most appropriately based on inter-dependence, rather than independence or dependence, so the links between communities are likely to be most fruitfully managed on the same basis. The need to negotiate with the wider world will exist on several levels. Firstly, and most simply, there will be an on-going need to discuss issues of mutual interest with neighbouring communities, issues such as where roads should go, the maintenance of utility lines that are held in common and other facilities that might be shared between two or more communities, in either the short or the long term.

Boundaries are important only in so far as they circumscribe the membership of the community at a particular point in time. At present, the layout of towns and cities does not lend itself to such divisions because of the need to move goods and workers quickly from one point to another, i.e. it is linear. In a community moving towards sufficiency this need to move around will be greatly reduced and, as a consequence, the clustering of houses and the way they are interconnected (as well as the means of transportation) may be very different, possibly circular and more all-embracing. The limits of a community may be real, physical lines of demarcation (as one house is naturally separated from the next by the functions that take place in each) but there may also be a blurring at the edges, with individuals having affinities with more than one community. Where people choose to live in a community may reflect their commitment to it.

Individuals with particular beliefs, interests and skills will always have the opportunity to negotiate a proportion of their transactions in what might be called 'virtual communities', groupings, potentially covering the whole globe, that exist only to further those activities. Some communities might choose to be nomadic; theatre troupes, for example, might move from venue to venue as they did in the Victorian era. And people will always have the right to move from one community to another. Fundamentally, however, for the *communal* dimension to be a living reality (and therefore contribute to each individual's quest for self-realisation) there must be clarity about both the physical limits that the community embraces and its membership. Ultimately, people are either in the community or not and, if they are, they should be actively contributing their time and energies to ensuring that commonly agreed goals are achieved. A fuzziness over boundaries and membership will indicate either a community that is growing or one that is declining.

The second area of ongoing contact between communities will be when certain services can be supplied only at a regional, national or even global level. Priority must be given to sufficiency at the local level and only when it is impractical for a particular service to be provided in this way should a larger framework be contemplated. Locally based, ecologically neutral sources of power, for example, may be the green dream but reality dictates that centralised production and distribution will be the only option for some time to come. As always, individual customers are in a weak position to challenge an institutionalised industry which has reduced its task and responsibilities to the simple formulae of profit and loss. Communities, however, are in a better position to change how things are done by negotiating contracts on behalf of *all* their members. Such negotiations might include an emphasis on developing more localised technologies (the basis of which exist in embryo but, because of the short-sightedness of the money economy, remain largely undeveloped), use of renewable energy sources, pollution control, safe working conditions, etc. Los Angeles (although hardly an example of a community based on freedom, equality and love!) has successfully imposed strict targets for automobile exhaust emissions, and manufacturers, faced with the prospect of their cars being excluded from the city, are responding.

Once control moves outside the community, power becomes diffused and the dangers of institutionalisation become ever present. Ways of counteracting this risk must be built into whatever structures emerge. As suggested earlier, a community is in a better negotiating position than a collection of individuals but there is no substitute for the two basic requirements of participation: being involved both in the practicalities of delivering the service and in the overall decision-making process. To achieve these twin goals, members of each community might be 'seconded' to the service so that not only is the community aware of – and therefore responsible for – how the service operates, but it will also be contributing to the 'cost' of the production – thereby reducing the overall 'price' to the community (such secondments might be particularly attractive to unattached young people who would enjoy the freedom and the possibilities it offered for new experiences and friendships).

By involving community members directly in the decision-making process, the tension between supply and demand is effectively brought back to the community to resolve, because different members – or groups of members – are responsible for making both the decisions that shape the service's overall direction and for negotiating the provision of that service to the community. The further away from the community in terms of both size and physical location such services or organisations are, the more difficult it becomes to maintain this tension in any meaningful sense and the more likely it becomes that individuals will cease to have a stake in what is being done on their behalf. It is precisely in those circumstances that impersonal, institutional forces begin to work towards ends other than the nurturing of human life and the welfare of the planet in general.

Provided there is a regular exchange of goods and services, transactions *between* communities can be conducted using a token economy that works in exactly the same way as the system *within* communities described above, i.e. one that is unique to those communities and based on an hour for an hour. If trading is irregular, infrequent or tending to be one way, however, there will be a need for a more universal medium of exchange: it is possible to conceive of a global currency that would allow, for example, a community in England to continue to enjoy tea from a community in Sri Lanka even though they might produce nothing that people in Sri Lanka might value. It is important to recognise that such a universal currency is not just the money economy under a different name. For a start there would be no need for currency per se, all transactions being conducted electronically through a centralised accounting system.

Secondly, like a true token economy, communities wouldn't have to be in credit to trade. In other words the 'currency' would be created as and when it was needed. Of course, communities that didn't have a regular trading arrangement might want to check out each other's trading history; if the 'purchasing' party was going ever deeper into commitment the 'selling' party might choose to trade with a more prudent community. One consequence of such a system would be the removal of the impulse to speculate or create wealth through usury. There would be no inflation and communities could enter into long-term agreements without the fear of economic factors beyond their control ruining what they were trying to achieve.

My community right or wrong
Just as individuals cannot grow and develop in isolation, so communities must develop ever closer ties with their neighbours, with their regions and, ultimately, with the global community if they are to realise themselves fully. The Janus-headed nature of hierarchy applies (see p. 45), reflecting the tension implicit in the polarity of independence – dependence and an acceptance that mature relationships are based on interdependence. If freedom, equality and love are to be living concepts, they must apply to relationships between communities as much as within them.

Any clearly identified group, neighbourhood or state has an interest in maintaining good relations with its neighbours if only to ensure that they do not attempt to appropriate what is not theirs. The same is true for a community. Achieving sufficiency at the expense of one's neighbours is short-sighted because, sooner or later, they will attempt to redress the balance. History continually affirms that there is no security in trying to hold on to what one has if it is at the expense of others who feel they have a claim on it. And without security there can be little scope for self-realisation. In this way, we all have a direct and immediate interest in the continuing negotiations that are necessary to ensure equality of opportunity for both ourselves and the other, individual to individual, community to community.

There is nothing inherently unacceptable about a community becoming 'richer' than others in terms of the quality of relationships available or having effective community structures that produce well-designed dwellings, elegant, durable goods, etc. What becomes indefensible is that these benefits should be enjoyed by depriving others of the means to do likewise. That is patently the situation in the world today where 6 per cent of the population (the so-called developed world) has direct and exclusive access to the best lands and uses 40 per cent of the available natural resources.[9]

The issue of ownership of land, and the natural resources (renewable, recyclable, non-renewable) that derive from it, is beyond the scope of a discussion that focuses on the *relationships* necessary to achieve a society based on community. 'Ownership' of 'land', and who has 'access' to it, will nevertheless be fundamental considerations in whether such relationships can be achieved in practice, both within communities and between them. For the purposes of the present argument it is enough to say that a community based on freedom, equality and love requires that each member (and ultimately all humankind) should, directly or indirectly, have access to land and natural resources to meet their basic needs of food, shelter and security and, beyond that, to allow them to realise themselves. In practice, the amount of land and natural resources each of us needs depends on who we are and how we choose to express ourselves. Someone whose talent is in growing food will need more physical space (land) than someone who excels in cooking it. Overall, however, it is possible to envisage what a notional community might require in terms of land and resources. How it is divided up in practice then becomes a matter for the community to decide.

Unfortunately, land and resources are not equitably distributed across the face of the globe and, as a consequence, negotiations between communities at local, regional and (ultimately) global levels will be necessary to address this basic imbalance if people are to feel secure. The idea that our notional community requires a certain amount of land and resources offers a starting point. A real community would then need more or less, depending on whether it had more or fewer members. If the community actually occupied more land in practice it might be expected to contribute to other communities via the universal currency system (and vice versa).

On this basis communities can elect to be as large or as small as they like (within the overall framework of a human scale) and even if they choose to live in inhospitable or marginal areas their right, directly or indirectly, to land and resources for self-realisation can be negotiated and any necessary adjustments made. Once the basis for this balancing has been established, boundaries of communities become no longer worth fighting over. If a group at the current boundary between two communities decides to move from one to the other, it may take physical land with it but it will also take its allocation of notional 'land' as well. If one exceeds the other a simple adjustment can be made. The same principles will apply to individuals or groups who wish to move to a distant community. In this way communities will grow and separate in an organic fashion that reflects their success in allowing individuals to realise themselves.

Just as openness will characterise relationships within a community, so it must *between* communities. Access to information (about, for example, population and land issues) is the only real basis for trust, and trust is fundamental if local, let alone regional and global, negotiations are to be successful. The ultimate prize, a world at peace with itself, is surely worth the effort of maintaining a free and open system of exchange.

Negotiations will inevitably depend on the qualities of the individuals involved and it will be important that the teams selected can demonstrate their commitment to the principles of freedom, equality and love. Just as mediators will emerge within a community whose skills are a reflection of their daily living, so the reputation of negotiators at all levels should rest on their wisdom in finding a way through problems, rather than on any ability to push through their own point of view. They will have to operate 'in community' with their opposite numbers. One of M. Scott Peck's more successful community-building exercises involved the Management and Trades Union negotiating teams from a national corporation.[10] So successful were they in doing a deal that was actually in everyone's best interests that they couldn't admit to the process by which the agreement had been reached; they feared that their constituencies wouldn't believe they'd done their jobs properly without some evidence that one or other side had been brought to its knees!

In the context of community relations, the processes of negotiation are likely to be subtle and take place over time, rather than through the instantaneous 'breakthroughs' much loved by the media. Time may well be spent discussing the values that are shared rather than the issue in dispute, because establishing consensus about principles minimises the dimension of personality and the need to assert oneself through having one's opinions accepted. As with all human transactions there can be no absolute solutions that are valid for all time. Individuals and communities will both win and lose. What is important is that a balance is achieved between the two and, more importantly, believed to have been achieved by all parties. That is the only basis for long-term equity.

Finally, the pursuit of self-realisation implies compassion for individuals and communities overwhelmed by loss or disaster. Any system of local, regional or global negotiations must include provision for aid in its widest sense. Humankind is but frail flesh in the face of the forces unleashed by Nature and, however successful we might become in living in harmony with the natural world, there will always be misfortunes that are beyond the capacity of any individual or community to withstand. In those circumstances the speed and generosity of our response will be a measure of our humanity.

6

The Sufficient Community
in Practice

Truth is first ridiculed, then opposed and, finally,
accepted as if it had been self-evident from the start

SCHOPENHAUER

After a necessarily impressionistic overview of some of the organisational issues facing a world based on freedom, equality and love, it is worth asking how, in more concrete terms, a sufficient community might satisfy even the basic needs of its members. The first two levels in Maslow's hierarchy can be summarised as the need for food, shelter and security. How far could a local economy meet these requirements and how might everyone be involved in the processes? To get an idea of what might be it is as well to start off with what is.

An apple a day

We so take for granted the piled shelves in our supermarkets that children might be forgiven for thinking that beans grow in tins. A quick look round the produce on display shows that no corner of the globe is unrepresented. Even the humble potato is likely to have started out in the Middle East. Surely, this abundance is one of the glories of the modern world: anything, grown anywhere, can end up, as if by magic, on our tables – bananas in the middle of winter, speciality cheeses and other produce from every part of Europe on never-ending display. We have overcome the seasons themselves, and borders that have been fought over for centuries have melted away.

And yet, more than half the world's population lives at starvation levels and, even within our own land, the lack of adequate resources to sustain a proper diet is the root cause of much ill health. Is the task to find the means to bring the cornucopia enjoyed by the few to the world at large? Or is starvation the inevitable attendant at the feast?

'Civilisation' means, literally, living in cities. And to live in cities requires an agricultural system that produces a surplus. Since medieval times the productivity of farmers in Europe has increased fiftyfold and, as a consequence, less than twenty per cent of the population now work on the land. During that same period, moreover, the population of Britain has grown from around 7 million to the present 56 million, an increase that has been possible in our already crowded island only because of an eightfold increase in yield per acre. And still we are a net importer of foodstuffs.

Our current level of population and preference for urban living, therefore, depends on the productivity of the farmer, the yield per acre and our ability to be able to import food from elsewhere. If any of these three factors were to fail for a significant period of time, catastrophe would certainly follow. That has been the experience of civilisations from the dawn of time. Several factors usually combine to make collapse inevitable. Climates may change. The population can, quite simply, outgrow the ability of the land to support it, leading to soil being overworked and a consequent falling away of yields. More often, there is a direct, perverse incentive in the overarching economic system that pushes the farmer to exploit the land for short-term advantage, ignoring the long term ecological damage that will follow – this is most obviously the case in the developing world where much of our imported food originates. Over a period of time an inability to feed people leads to civil strife, producing societies resembling that of the Ik described earlier.

In the developed world we appear to have achieved a steady state, a system of agriculture that keeps delivering the goods. Indeed, we are in a position where we routinely over-produce in certain areas, creating hefty surpluses that become an embarrassment and a joke. The price we pay for such continuing abundance is the increasing mechanisation of agriculture, producing the ever larger and apparently more efficient units needed to be economically viable. Farming has become big business, with short-term profitability inevitably determining the priorities. In simple terms that means ensuring that output price exceeds input cost but, in common with many processes in the money economy, that equation ignores factors on which long-term sustainability ultimately depends.

Producing food requires energy. Prior to the industrial revolution, when animals provided universal motive power, the amount of energy available from foodstuffs was frequently forty to fifty times that needed to produce it. Today, many sources of nutrition actually require more energy to produce than is yielded back as food. That kind of deficit farming, however profitable it might be, can be justified only when the energy put in comes from an infinitely sustainable source. Ultimately, of course, all energy comes from the Sun, but agri-business is heavily dependent on a particular form of it – hydrocarbon fuels, of which declining supplies will begin to have an effect on the Western lifestyle as early as 2025.[1]

Put at its bleakest, without fossil fuels there is no way that current productivity can be maintained by current methods and, with the uncertainty surrounding the impact of global warming on climate, it is a brave soul who looks to the next century with any optimism about the West's ability to feed its own population, let alone contributing to that of the world at large. Add to that the way in which we have been effectively divorced from what it means to produce food and it is possible to get a glimpse of the problem confronting us. We have become cut off from Nature in a way quite unique in human history. Even in the greatest cities of the past agriculture was an ever-present spectacle, with animals herded to market and a constant procession of people, wagons and barges bringing vegetables, grains and firewood – usually returning with the manure to put back on the land, land visible and accessible beyond the city walls. As a consequence, ordinary people understood the rudiments of cultivation and instinctively recognised its limitations. We are more like the nineteenth-century nobility who affected to have no interest in the land that brought them wealth and went to the wall uncomprehending as the embryonic world economy drove prices down.

How, then, do we set about reclaiming the land on which our futures depend? Firstly, we must be clear that it is in our interests to regain control of the processes that go to produce the food we eat. Once again, that means nothing less than becoming involved in those methods and having some ownership of them. Only in that way can *we* be clear about how sustainable they are and what *we* have to do about it if they aren't.

To get some idea of the implications it is worth reflecting that 'a standard suburban allotment garden (0.025 hectare) planted with vegetables and fruit can yield 2 tons of food per year from 350 hours of work (less than eight hours a week) at an energy ratio greater than 1.3. The allotment can yield 50 per cent more weight of food per unit area than a farmer growing heavily-fertilised and sprayed potatoes.'[2] Free range livestock could also be kept on larger areas of land. Using personal money a community could ensure maximum quality both now and in the future by encouraging those with a talent for working with nature to develop those skills, and by everyone adopting a lifestyle that minimised potential damage to the environment (access to and the use of water would be important considerations in this context, and the way we use this precious resource would have to be changed – sewerage systems, for example, might have to be adapted to minimise the use of water and maximise the fertiliser or energy potential inherent in human waste).

Such small-scale farming, with housing interspersed with crops, livestock and forestry is common enough in the developing world. It is also discernable much closer to home. The French, for example, have a practical attitude towards city gardens: it is not uncommon to see beautifully tended vegetable plots interwoven with flowers grown for purely decorative effect, creating a visual sense of wholeness. Communal farming and distribution is also springing up, especially in relation to the so-called whole foods. Spiritually based communities such as that at Findhorn on the Moray Firth have shown that communion with and a deeper understanding of the world's ecosystem may ultimately be more important in determining yield than a scientific, objective attempt to manipulate Nature for humankind's own ends.

In the epilogue to *The Ages of Gaia*, James Lovelock reflects on the present in relation to the aeons of earth-history he has been charting:

So why should I fret over the destruction of a countryside that is, at most, only a few thousand years old and soon to vanish again? I do it because the English countryside was a great work of art; as much a sacrament as the cathedrals, music, and poetry. It is not all gone yet, and I ask, is there no one prepared to let it survive long enough to illustrate a gentle relationship between humans and the land, a living example of how one small group of humans, for a brief spell, did it right?[3]

A personal view; but it would be nice if future generations could say the same about the fresh start that we, of these generations, must make if we are to have anything left to pass on to them.

The house that Jack built

We all need a roof over our heads. Apart from protection from the elements it should provide privacy and a sense of personal space that is free from invasion by others. Perhaps 'home' conveys its complexity as well as anything. It is, and has been, called by many names: a house, flat, hotel, boarding house, etc. and its physical appearance continues to vary according to fashion, climate and advances in technology, as well as the economic, political, demographic and social context. The factors considered necessary for a comfortable existence have also changed over the years so that our current aspirations might have looked princely, beggarly or simply irrelevant to our forefathers. Increasing specialisation has also meant that what actually takes place within the four walls of a 'home' has changed out of all recognition.

In short, different needs produce different kinds of domestic dwelling arrangement. Those needs change over the life cycle. Young people may prefer an openly communal lifestyle, implying flats, hostel or bed-sit accommodation. Families need more space and access to a safe, external play area, while the older person often looks for housing that takes account of reducing dexterity and mobility and that is easy to maintain. People with disabilities may require somewhere that is designed to meet their particular needs – which might include having a number of supporters passing through or staying overnight.

It is surely noteworthy that a society apparently committed to universal education and health services should be so strangely silent about extending the demand to food, shelter and warmth. It is the lack of these fundamentals that, on the one hand, stretch the health services to breaking point and, on the other, stultify any potential that education might liberate for the benefit of all. Communal health and well-being require that not only does everyone have access to housing that meets their needs but that a mix of dwelling places should be available so that no one should have to leave their community to find appropriate accommodation. Equally, particular types of housing should not be clustered together to form 'ghettos' in which individuals at particular stages of the life cycle or with particular needs are congregated and thereby lose touch with people older or younger than themselves. Friendships and access to the community network should be a matter of genuine choice. The current trend for people who have retired to move to isolated 'village'-type complexes represents both a withdrawal from wider society and a denial that old age is an inevitable and valued part of life.

The challenge for a community, then, is to provide a variety of good-quality housing that meets the needs of its members at a particular time. This imperative implies a degree of advance planning: changing needs are usually apparent some way ahead and it takes time to make the necessary adjustments to housing stock, including the possibility of new build. Looked at from this perspective, it is clear that housing is a community responsibility and that there is an inevitable tension between the needs of the individual and the community as a whole. How is the necessary balance best achieved? As always, it is helpful to take a brief look at how matters currently stand.

Britain emerged from the Second World war with 3.5 million homes damaged and 200,000 destroyed. 'In addition, it was thought that at least 750,000 homes were needed for new households created by post-war marriages and a rising birthrate.'[4] The priority was for numbers and speed. The result was the proliferation of high-rise estates that have become synonymous with the failure of local authority housing – of isolated estates beset by poverty, vandalism and unemployment. In fact, the numbers rehoused in this way were never great; 'it just *looks* as if there is more; the image of the modern city can be one of masses of high blocks .'[5]

The situation has hardly improved. There are currently some 100,000 homeless *families* (i.e., not including the single homeless) and a million homes that are deemed to be unfit for human habitation (out of a total of 22 million). One-third of our housing stock is at least 60 years old and 'nearly 90% of [houses] built between 1871 and 1918 are still in use. The situation is deteriorating.'[6] It is clear that neither state intervention nor the existence of a thriving private sector building trade has provided any answers to a state of permanent crisis. In fact, the type of housing being produced is largely institutional in character and lacking in any vision about what being human entails.

Fortunately, lessons have been learnt along the way. There is now a general recognition – if not a total commitment to following the message through – that housing is about people, and that means involving them in issues that matter in their daily lives. Community architecture is a growth industry, and there are many examples of partnerships between 'experts' and the people who have to live in the houses, producing results that not only mean better living conditions but also create a sense of community and a pride in what has been achieved that survives the departure of the professionals to their middle-class retreats. Housing associations and cooperatives have been particularly successful at renovating tenements and estates that had become bywords for all that is bad about today's city life. Crime and vandalism are reduced as people begin to feel a sense of communal ownership, of being part of something bigger than their own immediate desires and problems. Above all, it is an appropriate sense of scale that is the key to turning these situations round; and that scale is human.

By involving everyone, however indirectly, in the processes that sustain the meeting of basic requirements, the sufficient community takes these insights to their logical conclusion. The divorce between power and responsibility that is so apparent in today's institutional world – where experts wield great power but leave the ordinary man and woman to live with the consequences – is eradicated. Decisions relating to layout, design, scale, management and maintenance are everybody's responsibility and that recognition gives a great and liberating power to all concerned. There need be no bureaucracy, no cumbersome planning processes and no passive acceptance of what the experts think is best. The only obstacles to renovating and maintaining a community's entire housing stock, redesigning the layout to reflect a more communal nature or merely introducing improved standards of insulation, are the ability to find enough hands to do the work and someone with whom to trade 'universal currency' to pay for materials and/ or skills not available within the community. The development of good-quality accommodation for all need not be a pipe dream.

The notion of a sufficient community having control of all the processes needed to sustain life is a first step on a road that will lead, ultimately, to humankind being able to live in harmony with Nature. How that balance will look in practice is hard to imagine but, once again, James Lovelock offers a glimpse:

> it would involve the return to small, densely populated cities, never so big that the countryside was further than a walk or bus ride away. At least one-third of the land should revert to natural woodland and heath.... Some land would be open to people for recreation; but one-sixth, at least, should be 'derelict', private to wildlife only. Farming would be a mixture of intensive production where it is fit so to be, and small unsubsidised farms for those with the vocation for living in harmony with the land.[7]

It is a vision that would find support from the likes of William Morris and Raymond Unwin – the architect of the first garden city, Letchworth – both of whom recognised that beauty is essential for the human spirit. As Unwin said, 'we have become so used to living among surroundings in which beauty has little or no place, that we do not realise what a remarkable and unique feature the ugliness of modern life is'.[8] That was written in 1909, but the criticism is as valid today because our perception of beauty has something to do with scale, and human scale is as sadly lacking in today's cities as it was at the beginning of the century.

Beauty also means being surrounded by nature, allowing an interactive relationship rather than one that is essentially one-sided and, literally, outside one's self. For a community to be sufficient, therefore, implies not just the meeting of basic needs such as shelter but that those needs should be met in a way that enhances the quality of individual life. Even something as apparently mundane as housing should engage all our senses and should be built with the pride of the craft worker rather than the automation of the production line. To knowingly accept the second-rate in anything is to diminish our potential as human beings; it is to accept that we ourselves are second-rate.

Better safe than sorry

Security is as much a state of mind as a reflection of military might or sophisticated technology. One can still feel insecure in a house alarmed with every modern device including instantaneous access to the local police station. Indeed, it can be argued that the very fact that such measures are considered necessary *increases* one's sense of vulnerability, of unseen dangers lurking outside. Isolation from the wider world, whether as an individual or a nation, allows the imagination full rein, fuelling an innate paranoia until there are potential enemies around every corner. A society as fragmented as ours is thus fundamentally insecure when compared, for example, to a tribe living in the rain forests of Borneo, even though the latter's lifestyle might be considered more inherently uncertain and dangerous.

That increasing fragmentation also requires more and more 'glue' – in the form of agents of social control – to hold the pieces together. It has already been suggested that, in a highly institution-alised society, those agents will themselves become institutionalised and act according to agendas that have nothing to do with their stated aims. The example of the welfare monoliths, constructed to alleviate the effects of poverty, ill health and disability, has already been cited, but it is equally obvious in the police and criminal justice system.

In case after notorious case the evidence that formed the basis for conviction has been found to be 'unsafe', either because the scientific techniques used have been shown to be faulty or, more frequently, because the methods used by the police to gain confessions have been found to be suspect. Studies have shown that, often very early in an investigation, police officers form a view of the case and then set out to find the evidence to confirm it. They have an institutional goal – to secure a conviction (which is not the same thing as finding out who committed a crime) – and it is easy to see how, given the pressure to keep the ratio of solved to unsolved crimes respectable, or because of public outrage in a particular case, the tendency to cut corners can lead to outright abuse.

This tendency is even more apparent in set-piece confrontations with very differing groups of citizens. There have been many examples in recent years – from the 'hippy invasion' of Stonehenge (1985), to the London Poll Tax demonstration (1990), to the policing of animal rights protesters campaigns against the export of live animals (1995). A picture emerges of the police having decided *ahead of the incident* (and without reference to any independent judicial or democratic body) that the particular group poses a potential public threat. Officers are deployed according to carefully laid-out plans and, because the opposition has been defined as unlawful, all that follows becomes justified. In fact, the police frequently acted as if they were ambushing an unarmed band of some hated enemy rather than relating to a group of fellow citizens. Once a group is defined as being 'other', the mind set is totally indiscriminate, and the prospect of armed units patrolling the streets becomes yet another symptom of a gradual slide towards authoritarian rule in which the individual becomes expendable.

Justice – a prerequisite in a community based on freedom, equality and love – is about fairness, which is essentially a very human concept. It goes beyond a simple 'proof beyond reasonable doubt', which has become the province of the expert, to embrace notions of culpability, compassion and redress. The symbol of justice – a blindfolded female figure holding aloft a sword and a pair of scales – is a complex image. The scales show the importance of balance and symmetry, of not going to extremes in the pursuit of justice: the sword, the need for redress. The blindfold and the female figure suggest the male and female principles, and that one of the balances that has to be achieved is thus a synthesis of the paternal justice and the maternal mercy. A judgement becomes a weighty matter and, at its best, should be a creative response of head and heart that considers the circumstances of *all* the parties involved, including the victim.

Looked at in this light our current legal system is seen to be very one-sided, reflecting a need to protect the accused from arbitrary or summary justice. It is a principle that has not been won easily but the time has come to move on. In a sufficient community the very definition of crime would be different. Instead of measuring a de-personalised act against a set of rules to determine whether one of two options apply – conviction or acquittal – disputes between individuals would be talked through until *all* parties had agreed the nature of any injury and the course of action needed to remedy it or prevent it happening again. Ultimately, the whole community might be brought into the debate, much as the ancient Greeks used to listen to all sides of an argument before voting for action.

The issue of 'criminality' remains, however. Are felons born or created by circumstances? Do some individuals have a predisposition to commit crime or is all deviancy the challenge of the excluded? What part, for that matter, does evil play in human affairs? Nature and nurture appear as polar opposites in this context, and the way forward is through the creative and on-going resolution of the tension between them. Selfishness has always been a spur to action and is necessary for survival and personal development alike but, within a value system that doesn't actively encourage its expression, we are more likely to achieve a healthy balance between the self and the other within our selves.

Each 'offence' becomes a new challenge to the established order and, on the basis that we are also seeking a balance between order and disorder (not the maintenance of order at all costs), should be valued for the lessons that might be learned about individual and social nature. As well as individuals *having* to change – and sanctions being imposed if they don't – it may be that the community will have to adapt to satisfy needs hitherto unmet.

People would be valued for being trustworthy and open in their dealings with others. The quality of the relationships within the community would be of prime importance, and part of that focus would be found in the creation of mechanisms for effective negotiation of differences – which underlie most of the cases currently dealt with under civil law. They are essentially issues between *people* and concern human expectations of, and perceptions about, one another. The adversarial approach employed by the current legal system is singularly ill-suited for resolving such matters because it produces winners and losers. Establishing a mutual understanding of what has happened, and why, should be the aim. Only then can reconciliation occur, including appropriate restitution (possibly on both sides) and a commitment to preventing a recurrence of such difficulties.

Offences against the community would have to be handled in a different way, although, assuming that basic needs are being met and there are genuine opportunities for self-realisation, such incidents are likely to be rare. As already suggested, vandalism would be greatly reduced, if not entirely eradicated, just by giving people ownership of their community. More likely would be a failure to discharge communal responsibilities, and systems – which would have to be accountable to the whole community – would have to be devised to 'hear' such cases and decide on appropriate restitution. Expulsion from the community would be the ultimate penalty in prescribed instances but, at all stages, the individual should have the right of access to mediators and/or advocates.

As a postscript to this section on security it is worth reminding ourselves how much we have come to rely upon experts and the effect that this handing over of control has on our daily lives. From accountants who organise our lives in tax-efficient ways to insurance advisers who help us think the unthinkable and anticipate our demise, there are people eager to shape and advise us. Environmental health officers monitor restaurants, imposing requirements that the average domestic kitchen would fail. Planning, fire and building regulations all have to be adhered to in public buildings, ensuring safety but at the price of creating environments that can be institutional in appearance and difficult to negotiate for people with disabilities. Our dependence on experts is best illustrated by the increasing inclination to try and sue when something goes wrong – encouraged by the burgeoning pack of legal advisers who flutter around, picking over life's daily tragedies: it must be some-body's fault, somebody should have prevented it happening.

Power involves the acceptance of responsibility and, if *we* are to regain control of our lives, that means accepting responsibility for the choices we make. It can be argued that the armies of regulators are essential for our welfare in a world that seeks to exploit us, but that perception stems from our fragmented, role-dominated society. The experts attempt to compensate for our inability to take respon-sibility for such basic safeguards ourselves and, in so doing, impose rules that do not reflect the lives of real human beings. When we visit friends we do not ask to inspect their kitchen or check whether there are sufficient toilets for the number of people involved; in so far as we make judgements on these issues it is on the basis of what we know and observe about them as people. In a sufficient community, not only will we be much more intimately involved with the processes that are going on around us – whether that be in terms of food preparation or building work – but the community will have an interest in ensuring that we are aware of health and safety issues that might have an impact on the decisions we take on a daily basis. Expertise is like authority: it should be vested in individuals, not roles. We should accept someone's advice because we know and respect them as an individual, not because they operate within a prescribed legal or professional framework that gives them the right to impose their views on others. At the end of the day, we cannot grow and develop without taking risks. A society that routinely gives responsibility for risk management to others is inherently unhealthy and thereby unsafe.

A stitch in time

When we look at meeting even basic needs we inevitably assume the wide range of artefacts that are considered necessary to satisfy them in a dignified and valued way – furniture, pots and pans, clothing, etc. For people to be involved in the processes that create the things they use inevitably raises questions about the scale of production methods and how far 'efficiency' dictates that certain operations are done at local, regional or even global levels.

For example, is it a 'better' table if I have been involved in the process from the growth of the tree to its final incarnation as a piece of furniture, or if I am the drawer-handle-putter-onner in a production line that may stretch across continents and be largely computer-controlled? And what does it mean if one table is 'cheaper' than the other? This debate is universally resolved today in the market place through the medium of supply and demand which, by its very nature, is likely to put a premium on price. From Adam Smith onwards the philosophy of producing more at a lower price has predominated and, as a result, we have got what we have been given which, in turn, is a reflection of what manufacturers think we want or think we can be persuaded to buy. In that sense our 'demand' is not a measure of us as individuals but a generalised perception of what we, the masses, might want.

In a sufficient community, demand is likely to emphasise difference rather than more of the same. That shift will require 'consumers' and 'producers' to come together and negotiate a finished article that satisfies both their needs. A table is a table, but my table will be different from yours because we are different people and we have created different environments in which we wish to place our tables. Although such face-to-face contact is the ideal, it is clearly impracticable for small-scale communities to be totally self-sufficient, if only because no place on the globe has access to all the material resources needed to meet the needs of all its members. Equally, there are any number of items where 'individualised' production is not appropriate – from pins to piping, from paper to paint. They are essentially utilitarian in character and can be turned out in any reasonably equipped workshop anywhere in the world. Rather than finished products moving around, it therefore makes more sense for primary or rough-worked materials and components to arrive in communities and to utilise the infrastructure available there to produce a finished good that reflects local and/or individual needs. Specialised machinery might also travel round several communities and thus be available on a much wider basis than if it were static.

Some processes currently have an optimum size and location by virtue of the quantities that have to be handled to produce a given result, or because they have to be sited next to natural resources that occur only in certain areas. It is difficult to envisage, for example, the production of steel being reduced to a scale where it would be viable for each community to have its own plant. Equally, the research and development work required in relation to electronics – on which much of the communication systems within each community will depend – would continue in specialised communities (although the assembly of finished products from components could be done locally). Similarly, the fabrication, launching and maintenance of satellites will become a regional, if not global, operation. In the same way transport systems, such as railways, that allowed the transfer of people and goods could be serviced by communities strategically located along the route. To ensure that these processes do not become ends in themselves, it is crucial that the people involved are committed to the principles of freedom, equality and love and can demonstrate their willingness to live 'in community' to the outside world. In that way, other communities, unable to be involved, can be assured that the products are life-enhancing.

The debiting and crediting associated with such arrangements would be handled through the universal money system because the essential feature of personal money is that it is restricted to transactions within a community. As has already been suggested, all such transactions would be managed electronically; there would be no need for the exchange of physical money. Each community would have access to a global network, the administration and servicing of which would be done by specialised communities.

The central imperative of the money economy – that everything has a price – leads people to accept the mindless repetition of the production line because the wages are good or because they have no alternatives. If that sense of being unable to live without money were replaced by a system of personal money, through which the basic needs of life are guaranteed by the free exchange of time within a community, what would the impact be on 'work'? The pursuit of self-realisation suggests several pay-offs that should stem from any activity. Firstly, there should be a sense of competence and achievement that enhances self-esteem, a feeling of having used one's skills and judgement to good effect. Secondly, the activity should produce a result that is valued both by oneself and other people. Thirdly, it should foster a sense of belonging to something beyond oneself (e.g. that one is freely doing something for someone else or actively engaging with others in the accomplishment of the activity). The more these three features are evident, the more the individual is likely to respond and seek to improve their performance. By contrast, a lack of such rewards will lead to boredom, disengagement and a loss of self-worth. Paradoxically, in many activities, the more competent and experienced one becomes the more one takes mastery for granted and the less one's skills are valued; which is why it is important for individuals to be able to expand their horizons continually by becoming involved in a range of activities rather than being tied to a single 'job'.

To promote self-realisation means encouraging people to use their whole beings, to avoid specialisation unless this is an expression of themselves. The whole thrust of technological advance runs counter to this requirement, breaking tasks up into ever more specialised activities and seeing the human contribution as fitting into the logic of the process. There is a need, therefore, to develop *appropriate* technology that assists each individual to produce something that has utilised the maximum of their potential as a human being. The word 'appropriate' does not imply a return to techniques that predate the industrial revolution or the adoption of a middle class arts and crafts philosophy. In *Small is Beautiful*,[9] E. F. Schumacher calculated that only 3.5 per cent of Britain's total time was spent in formal production and that the trend was to reduce this proportion still further. 'Imagine,' he goes on to say,

> we set ourselves a goal in the opposite direction – to increase it sixfold, to about twenty per cent, so that twenty per cent of total social time would be used for actually producing things, employing hands and brains and, naturally, excellent tools. An incredible thought! ... There would be six times as much time for any piece of work we chose to undertake – enough to make a really good job of it, to enjoy oneself, to produce real quality, even to make things beautiful'.[10]

By definition, such labour-'inefficient' modes of production lend themselves to small-scale, decentralised settings. Individual communities could support a variety of such production units and allow individuals with entrepreneurial skills (the ability to first visualise and then create something that didn't exist before) to express themselves for the benefit of all. Labour would be available within the community and people would have a free choice over whether they participated or not (because the meeting of basic needs would not depend on earning a wage) and could negotiate conditions, hours, environmental factors, etc. More likely would be partnership arrangements between a number of individuals.

Give me a child to the age of seven

Our future lies with our children – quite literally. If we are to grow old in a community that values us and is prepared to meet our increasing needs in a way that leaves us in control, it will be our children who make it a possibility. Their ability and commitment to sustain that environment will be determined, in no small measure, by the world we have created for them and, specifically, through the experiences they have had as they grow to adulthood and assume their full place in society. The education – in its broadest sense – that we have provided will be a good indicator of what the future might hold. It will also give us clues as to how we will have to change our child-rearing patterns if we are to move in the direction of the sufficient community.

Unfortunately for us, the system to which we entrust our children for ten to fifteen years is facilitating the advance of the Unholy Alliance, rather than in any way frustrating it. When upwards of a thousand individuals are herded together day in and day out, divided up by age or sex, and expected to comply with a set of behaviours that bear little relation to how people act in the outside world, it is impossible to avoid the effects of institutionalisation. Pupils emerge comfortable with the constraints of organisational living, moving into a job as easily as trying on a new suit of clothes. In time, both will feel as if they had been made for us: they become a statement of who we are. In reality, of course, work is not part of us, we are part of it, and we have become tailored to its requirements rather than the other way round. Habit ('of all the plants of human growth, the one that has least need of nutritional soil in order to live'[11] has taken over, and the seeds of that diminution of life's possibilities are being sown in our schools.

The attempt to measure performance – through universal examinations or rankings in class – reinforces a basic belief that all things can be measured. It is but a short step to all things having a price. Learning is segmented, knowledge compartmentalised, reflecting an increasingly fragmented nature of society at large and contrasting with the greater 'wholeness' of simpler societies in which knowledge is generally shared by all its members and is acquired not in a specific institution but as a natural part of the life cycle.

The responsibility laid on the individual is not self-improvement or self-analysis but achievement within the parameters laid down by the system. In that sense it mirrors the view of personal responsibility fostered by the money economy, namely to accumulate as much wealth as possible by striking the best deals one can. Interactions become depersonalised through the medium of exam results that, like money, can be traded on the job market. No thought need be given to the 'failures' who have left school or been moved to different groupings, just as no thought need be given to those at the bottom of a society stratified by success. In fact, children learn co-operation in an overtly competitive environment; it becomes a means to an end rather than an end in itself.

Stockpiling a set of pre-packaged modules does not equip an individual to confront the major questions of their existence, such as who they are and what purpose their life might have. Nor does it encourage an awareness of the beauty and mystery of life. By placing the emphasis on the acquisition, rather than the ordering, of knowledge, the education system has set itself firmly in the relativist camp, failing either to differentiate between levels and types of knowledge or to provide any unifying concepts by which such an evaluation might be undertaken. It is an environment in which the objective and reductive tenets of traditional science can be absorbed unquestioningly, thereby providing a lifetime's legitimisation of materialism and the need for continuing 'progress'.

It is a paradox that education, at once the most universal and potentially nurturing of all the so-called caring professions, should have become the foundation on which the logic of the money economy – that most impersonal of forces – should rest. The imperative of economic growth has truly distorted every aspect of human existence. The pressures that apply to the corporate world of business are now being applied to the health, education and social services because there is no longer any distinction between their aims and their meaning to human existence. Economy, efficiency and effectiveness are everything and that means defining your product – attainments in education, throughput and mortality figures in health care – and focusing ever more narrowly on demonstrating that you are producing value for money.

That situation is going to get worse. Teachers are increasingly finding themselves at odds with the system in which they have to operate. Many do manage to encourage children to learn in the traditional sense of exploring and mastering their environment. Some can still convey a sense of awe and wonder in relation to their subject that will stay with their pupils for life. All that demonstrates is that the human spirit will reach for the light no matter how depressing or discouraging the circumstances: but they are small rearguard actions in the face of an institutional onslaught. The reality is that we are preparing our children for a specialised, partial and dehumanised existence. If we are to encourage a personal journey towards self-realisation within the context of a community based on freedom, equality and love we must look again at the process of education and what it is trying to achieve.

Although much of the information being imparted will be common, each child will be encouraged to chart its own course, sometimes on its own, sometimes co-operating with others. There will be a greater emphasis on the linkages between different aspects of knowledge – rather than on their separation into clear-cut subject areas – and on the nature of knowledge itself. Co-operation between children will be important – possibly across, rather than within, age groupings – and there should be an early experience of negotiating about issues that affect them. Achievement will be in terms of an increasing acceptance of responsibility within the community and a growing awareness of the potential within oneself. As always, mechanisms would be evolved to ensure that individual children, their parents, teachers and the wider community would be involved in, and share responsibility for, what is happening.

Education is, at heart, an exchange between the teacher and the taught and, as such, is uniquely suited to the kind of token economy on which a sufficient community is based. If I, as an adult, wish to learn a skill or gain insight into a particular area of knowledge, I could negotiate an arrangement with someone I perceive to have the wherewithal to help me learn, and the kind of personality I am likely to respond to. The 'teacher', in turn, has the right to consider on what basis, if at all, they will respond to my need for instruction: am I truly committed to learning, how many similar requests they can meet, etc.? People who are particularly good at certain things will become known and sought after and may, as a consequence, choose to devote increasing amounts of their time to teaching, thereby giving themselves space to develop in other ways. It is a process through which natural differentiation occurs, with individuals increasingly going their own way at their own pace.

Such scenarios are not uncommon in adult education, and there is no reason why the same principles should not be applied to education generally. To go down this road, however, implies smaller groupings in more normal surroundings. Teachers operating from their own homes and taking children from their immediate neighbourhood offers a possible solution, but it is a model that raises a number of questions. Although parents (and their children) could choose from a number of options, who decides what is taught, how are standards monitored and how is the competence of teachers to be judged?

From questions such as these arise the whole panoply of professional expertise and yet they are, at heart, very simple questions. Most parents can articulate what their expectations of education are and have a sense of how their child is progressing. Equally, one does not have to have a Ph.D. in Educational Theory to be able to participate in a community debate about the nature and scope of education. A basic framework could be agreed 'in community' and the task of progressing it given to a small group to discuss with those involved in day-to-day teaching. As with all transactions the emphasis would be on clarifying expectations through on-going negotiation. Parents, children and the community as a whole would be much more involved in the process of education and teachers would not feel so isolated and unsupported.

The issue of competence is complex. What makes an individual competent to undertake a particular task? A degree or diploma in teaching merely indicates that, at some point in the past, the individual was sufficiently committed to pursue formal training. It says nothing about that individual now or the quality of the training. Competence should be a statement about ourselves; a creative, living expression of our being. From such a perspective our ability to succeed is a source of satisfaction, self-esteem and self-renewal, gaining us respect from others and, as a consequence, encouraging us to continue and improve our performance. It is a transaction between two people in the here and now rather than something that is conferred by a distant expert.

To maintain the vital spark of commitment in any activity requires a network of whole relationships (rather than a series of roles). Having fewer children means there is a greater opportunity to share with parents the joys and tribulations of the day. Parents themselves may choose to spend part of the day actively involved in what is going on. Equally, other interested adults might become regular helpers. The task becomes a partnership with all parties working towards the same end.

There remains a need for what might be called peer group support, sharing with others engaged in the same task, exchanging ideas and exploring solutions to mutual problems. Modern technology offers almost universal opportunities for exchange in both didactic terms (an individual or group presenting their approach on particular issues) and interactional terms (through phone-ins, etc.). Teachers could form their own virtual communities, using their own unit of exchange, or contribute through the universal currency system to local, regional or even global bodies who would offer such services – including buying in programme material or experiences from teachers themselves. Because such ventures are both relatively cheap to set up and do not require large numbers of people, new ideas could be launched with ease.

It was suggested earlier that one way to counteract the institutional tendencies of today's schools would be for teachers to take children into their own homes. Communities may equally decide to provide purpose-built accommodation which nevertheless remained on a human scale. There are many occasions when a domestic house is inappropriate or impracticable. Games require many players as well as pitches and equipment. Resources such as computers, facilities for music and drama, will all be available within the community and should be accessible to everyone, once again precluding the need for a specialised, educational campus. Craft and technology could be learned in workshops and industrial units run by members of the community (with older children able to use credits they had earned to extend their choice in these vital areas). In this way education becomes a whole-life experience, directly linked to the day-to-day experiences of the community, rather than separated from it. Time and organisation are needed to create the network of opportunities described above, but half the problems of life today occur because there isn't the time (or the breadth of relationships) to consult the range of people that would allow such flexibility to be built into the curriculum. In a community based on negotiation there will be very different rhythms that will, ultimately, be time- and labour-saving.

The approach outlined above changes our concept of childhood from a static, institutionally dominated condition – in which individuality sprouts in spite of the system – to an open, self-generating potential for being, in which the relationship with the adult world is in a continual state of re-negotiation. As an inevitable consequence, growing up implies the acceptance of increasing responsibility for the self and other. It is the only basis on which self-realisation can be sustained in the present as well as the future.

A further benefit of this child-centred approach is that many more adults are likely to become involved with each child, over a much longer period, than is currently the case. In a very real sense the responsibility for children will be communal rather than being left solely to the parents and a few detached individuals who have nothing to do with the child outside school. Together with the networks that parents will develop as part of their own self-realisation, this access to wider communal support will remove many of the pressures that our fragmented society places on the relationships of parents and, perhaps, create more realistic expectations about the long-term commitment needed to raise children. The emphasis on using one's time effectively rather than having to fit caring for children around a full-time job would also benefit parents, allowing them to devote time to their children while at the same time continuing to develop their talents through the use of personal money. It would no longer be a choice between parenthood and work.

Health of nations

Self-realisation implies a concern for the well-being of others and a genuine desire for their development towards wholeness. It is that absence of wholeness – literally the self divided against itself – that contributes to a range of ailments from feeling run down to a variety of stress-related illnesses. Having a sense of one's own integrity and of being somebody within a community of significant others is a basic requirement of physical, social and spiritual health.

Within this context, pain and suffering are not necessarily something to be smoothed away with the aid of a pill. True personal growth often comes through crisis and the resolution of conflict within oneself. Life is a vale of tears but from that inescapable hurt come the qualities that make us most human: courage, generosity, loyalty, humour, humility, perseverance, etc. The tragedy is that most pain we experience today does not come from within but is imposed from without by the conditions under which we live. Arbitrary it may be, but malice is not implicit in the institutional world. Once an individual has been strapped to the rack, however, there can be no personal resolution to the pressures they are forced to endure; in the circumstances it is hardly surprising that many are broken or seek solace through anaesthesia of one form or another. We call it the tension of modern living.

Once basic needs are being met, the qualities most likely to be in demand in a sufficient community will be what are loosely called 'counselling' skills. It is often said that many people approach their doctor not because they are feeling ill but because they want someone to listen to them – to *them*, to let them occupy centre stage in a world where they are essentially walk-on players. Of course, doctors do not have the time for such frivolities and dispense a variety of pills instead, which produce their own problems. Effective listening takes time, but time is only an issue when one is forever pursuing wealth. When an hour for an hour is exchanged within a personal money system, one's day is suddenly one's own again, because time itself provides the limit to what can be achieved.

By meeting basic needs and providing both purpose to individual lives and the means to resolve the inevitable crises that living throws up, the community can 'cure' most of the illnesses that currently pass through a doctor's surgery. There remains, however, a range of maladies that cannot be tackled solely in this way. These include physical traumas caused by accidents, diseases brought on by environmental factors, epidemic diseases, etc. The present way of dealing with these potentially life-threatening events is through the treatment of specific symptoms. Western medicine is firmly rooted in the scientific tradition which essentially sees the human being as a vastly complicated machine. Understanding the body relies, firstly, on sub-dividing it – as one might divide up the fuel and steering systems of a car – and, secondly, on studying cause and effect to identify what changes in the observed symptomatology a specified intervention brings. The existing model of the illness is then confirmed or altered accordingly. This process produces an image of sickness as being separate from any actual human being. Real people are being compared to the model and, as a consequence, their unique individuality is seen as being incidental. It is an approach that emphasises doing things *to* people – in the way one would service a car – rather than involving them in restoring *their* health.

Other, older traditions stand this approach on its head and start with the person who is unwell, working outwards until a unique picture of personality, symptoms, circumstances, etc. has been established and an individual cure prescribed. Two people who are diagnosed as having a similar complaint might, therefore, undergo different treatments. It is a convention that requires a deep understanding not only of illness but also of human nature and is, therefore, almost wholly dependent on the skills of the practitioner. When it works, it enhances both healed and healer but there is, inevitably, a margin for human error. By contrast, technological medicine is like using a sledgehammer to crack a nut. In theory, any competent doctor should be able to diagnose the objective illness and prescribe the necessary treatment The patient is either cured or not. Success confirms the confidence and credibility of the medical profession; failure is invariably seen as poor diagnostic skills or an incomplete picture of the disease that will, in time, be filled in by the scientific method.

As with scientific application generally, its very success blinds us to its inherent limitations and its institutional consequences. The medical view of the human body as an object has led directly to the growth of hospitals as the setting where serious medicine is conducted. Hospitals are places where objects (people) are processed according to set routines (treatment) to create a standardised, finished product (health). As in any factory, the race is on to find ever more efficient – and labour-saving – methods, including computer technology and robotics. Individual human beings were never further from consideration.

The fact that Western medicine and alternative therapies are so obviously on divergent courses would suggest that we are facing polar opposites and that a new balance is needed. The opposites are faith and reason. Reason tells us that the human body can indeed be reduced to an objective reality that can be manipulated in the same way as any other matter on Earth. Faith tells us that the human experience is more than the sum total of those material elements.

The challenge is to find a way of treating ill health that takes something from the two traditions. Starting from the premise of a community founded on freedom, equality and love, it is easy to see that there should be more knowledge and understanding of each individual and more commitment to their well-being than is currently the case. That is an important prerequisite for us to keep control of *our* illness. If we also have an ongoing ability to negotiate about our circumstances within a context that allows us to make realistic choices, we are in the best possible position to make the decisions that will effect our well-being. One of the major choices we face is whether to be treated at home. In practice, however, that choice is effectively ruled out by the lack of community provision in terms both of personnel and equipment. Many people may prefer to go somewhere special for their treatment but, by having that choice denied them, they, as much as those who might prefer to remain at home, are surrendering themselves to a process that takes them over.

Hospitals are where the power and professional prestige lie, and it is hardly surprising that the allocation of health funds reflects that position. There is nothing inherently feeble about nursing techniques or technology, however, that rules out treatment in the home for most conditions. Dentists, for example, have traditionally operated from community settings – often their own homes – and doctors now do many routine operations in local health centres which could provide a launching pad for a much greater devolvement and demystification of the medical machine. If the purpose of the intervention encourages human interaction, so will the technology; equally, if the goal is abstract research we shouldn't be too surprised when the machinery isn't user-friendly. By the same token, a piece of equipment that is so expensive or so sensitive that it needs to be housed in a special building is more likely to require people to fit in with it than the other way round. If we can shape technology by being clear that the aim is always to promote *human* values (as determined in each sufficient community), there are plenty of people in the community who have the skills (and, in many instances, the training) to contribute to community health through a personal money economy, allowing people receiving treatment to have real choice about who comes to visit them.

Accident and emergency facilities will continue to be needed, and some specialisms may require more centralised resources either in terms of expensive equipment or because numbers are too small to justify a local response. Where a community is unable to meet its own health needs – and this basic need has to be satisfied before higher-level needs can be met – services covering several communities might be necessary, and the principles governing the links between communities – see Chapter 5 – will apply.

This debate raises the issue of specialisation in a sufficient community. The pursuit of self-realisation implies an opening up to a range of experiences that will expand and develop the individual's range of talents and level of maturity. It contrasts with the closing down that occurs when institutionalised roles come to dominate; people become one-sided, no longer experiencing themselves and losing touch with who they are. Specialisation is one of the more obvious symptoms of institutional life, but it is differentiation based on task not talent. It is a process of increasing compartmentalisation that leads away from wholeness and integrity. By contrast, true specialists – people who focus their talent in a particular direction – are channelling their whole being into the enterprise rather than having to cut parts of themselves off.

All of life's experience is grist to the specialist's mill, as many famous artists and scientists down the ages have shown, whether it be in particular forms of surgery, the development of one branch of technology or an exploration of a single art form. Specialisation doesn't therefore have to run counter to the principles of freedom, equality and love provided, firstly, the choice is not made at the expense of someone else's potential for self-realisation (as much of current medical practice is because it effectively denies the other) and, secondly, the individual is making the choice as a positive expression of, and not as a means of hiding from, themselves.

Health and welfare, like the meeting of all the other basic needs we have described, is essentially a simple transaction between people. The use of a personal money economy offers the opportunity to break the mould of increasing institutional dominance and to re-assess what we really need and want from life. At the end of the day, however, the quality of that life and its durability depends on the integrity of those around us. A system that allows us to exchange skills and needs at a local level is no quick-fix solution. Initially at least, it will be harder work than any of us can possibly imagine, but the longer we leave it the harder it will become. Only an ever-deepening exploration of the potential within each of us and the communities we create can offer us the wisdom to find our collective way through the maze that is life. One day, hopefully, that journey will reveal the full inadequacies of the mind map that we are calling 'community'. And then it will be time to start the process all over again...

7

A Summing Up

It is only by thinking things out as one lives them, and
living things out as one thinks them, that a man or a
society can really be said to think or even to live at all.

PATRICK GEDDES

All new movements start with a vision and, today, many people
share a vision of living in community with their neighbours
and in harmony with the natural world. At present that vision is
fragmented, incoherent and partial, existing *within* people or small
groups rather than being external and available to all. The challenge
is to create a language through which the vision can take wing and
become generally accessible; through which people from Norwich
to Newfoundland and Nairobi to Nepal can begin the long process
of translating the vision into reality. The concept of the Sufficient
Community is offered as one step along that road. It is a beginning
not an end. The end will be achieved only when it is time for
another vision to be born.

At its simplest we are seeking to create communities that are
essentially human in scale and value.

To understand the implications of this statement we must remind
ourselves what it is to be human. We all have needs, ranging from
the simplest (food and shelter) to the more complex (esteem and
self-realisation). Except for the most basic, these needs are usually
met through the links we have with others. The more complex
our needs become, the more we need significant others through
whom we can explore and express our developing selves. It is
through such networks of mutually supportive relationships that
individual self-realisation becomes a possibility. A basic definition
of a community is therefore a collection of such networks, and it

prospers when all its members have the opportunity to explore and express themselves in ways that encourage maximum personal growth and development. Strength lies in diversity and the ability of each to contribute their own unique perspective.

We also explore the potential within ourselves when we take decisions. For much of the time we are unaware we are making an unending succession of choices and, as such, are in a continual process of negotiation with those around us and with the external environment in general. Mutuality and equity in negotiation are central to the kind of relationships described above.

This vision of what it is to be human can be encapsulated in the familiar phrase 'freedom, equality and fraternity'; the freedom to become oneself, a freedom that should be extended to all in equal measure, and fraternity (or 'love', to give it a more all-embracing title) that offers the possibility of reconciling the opposition inherent in the concepts of freedom and equality by recognising that to become oneself depends fundamentally on others being able to realise themselves. It is a virtuous circle: to truly empower ourselves we have to empower others.

Of course, a community is more than the sum total of the individual transactions that occur within it and, once it embraces the desire to be in any way self-governing or self-reliant, co-operative structures become necessary. Many activities are communal in nature (e.g. road building, disposal of rubbish, maintenance of libraries, etc.) and, although undertaken by individuals, require a community mandate to define how they should be done. There are decisions to be made about how to support individuals who are not fully in control of their lives by virtue of ill health and disability. Communities will also have to negotiate relationships with surrounding communities, regions and, ultimately, with the global community itself.

To organise such activities, communities require decision-making structures, and the power relations that will inevitably develop must reflect the basic aim of maintaining human scale and value, of ensuring equity in negotiation, etc. Recognising how important ongoing participation is for self-realisation and, by definition, for the health of the community, implies that communal structures must embrace the requirement that everyone should play an active part in the decision-making process. Ways of meaningful involvement need to be explored. Ultimately, it will be equally important that everyone is involved, however indirectly, in the social and economic processes that take place within the community.

Structures of decision-making must also take account of the nature of reality. Most issues facing individuals and communities can be termed divergent in the sense that they tend to produce a polarisation of views. Thus, the problem of traffic congestion in our cities can produce passionate exponents of both the laissez-faire – do nothing and let market forces price cars off the streets – and the interventionist varieties – ban cars altogether and create efficient public transport systems. In practice, policy becomes a complex synthesis of both which inevitably produces its own clusters of polarised problems. (Compare this dynamic with the convergence typical of the scientific method, where argument tends to converge towards a single agreed solution which becomes the orthodoxy through which the future is explored. It is one of the legacies and successes of the scientific method that we have come to believe that *all* problems should be solvable in this way.)

The essentially dualistic nature of existence means that we are in a continual and creative attempt to reconcile opposing elements both within ourselves and within the environment we inhabit (e.g. maleness/femaleness, rightness/wrongness, competition/co-operation, etc.). From this standpoint there can be no absolute or 'right' solutions to problems, only 'outcomes of the moment' – more or less successful attempts to find balance and harmony – which depend on the wisdom and integrity of those involved. Although it will be appropriate for many problems to be handled by individuals, increasingly groups working 'in community' will be the best way of responding to the complexities of life. Being 'in community' implies emptying our selves of preconceptions, prejudices and a desire to convert others to our point of view. It is an environment in which all points of view can be heard and weighed before consensus emerges. It is a consensus that needs to be constantly reaffirmed if concepts such as human, community, self-realisation, negotiation, empowerment, self-reliance, involvement, etc. are to be translated into daily reality. Ideas and structures that remain unchallenged begin to take on a life of their own that will subvert, and ultimately overturn, the values that the individuals they embrace are trying to implement.

To follow such an agenda will not be easy, but what have we got to lose? For a few it may mean losing wealth and power, but in return will come true security, the kind of security that cannot be bought with money or position. For most of us it will mean losing a way of thinking that requires very little effort, but it will also mean losing our isolation and the pervading sense of purposelessness that characterises life today. In the process we may discover our true selves.

The challenge is whether individuals who have become detached, institutionalised and morally blind can grope their way towards the light. Grasping the fact that we must all make a stand on the basis of the values that we hold dear is a first step. The values implicit in freedom, equality and love embrace all the rights that have traditionally been accorded to the individual and provide a rallying point for people who wish to regain control of their lives.

It won't come easily, and the Unholy Alliance will fight every step of the way, but the experience of Eastern Europe suggests what can be achieved. Once a people ceases to believe in the necessity of a particular state of affairs it is but a short step to sweeping it away. Can we not find the resolve to confront what is, on the face of it, a more benign dictator?

Prospero, at the end of *The Tempest,* casts off his magical powers and faces life as himself:

> Now my charms are all o'erthrown
> And what strength I have's mine own.

We have to dispense with the magic of the Unholy Alliance and face the reality of a world in which our every action is accountable. The promise that 'community' offers – and that goes some way to justifying the hard work and sacrifice involved – is that we will be able to live with ourselves, our neighbours and our environment in a harmony not yet achieved on this planet. The alternative stares at us from the ruins of Oradour-sur-Glane.

Epilogue:
Giving Love a Chance

In my end is my beginning
T. S. ELIOT

It is natural to wonder what the future holds. Who amongst us would refuse the chance to travel through time and see how people are living in a hundred or a thousand years from now? – if indeed they are still living, because we oscillate between optimism that life *is* going to get better and pessimism that we are doomed to self-extinction. Such fluctuations reflect how we feel about ourselves as much as external reality and it is fortunate that the life-affirming side of human nature predominates (we could hardly have survived as a species if it didn't). It is that spirit that will have to be engaged in each of us if we are to find a way forward. The question is 'how?' You may well have been thinking that it's fine to *talk* about building communities based on freedom, equality and love, but where does one start? Your priorities at the moment are probably keeping a roof over your head and paying the bills. What time do you have to change the world?

If, by changing the world, you envisage a new order in which all the contradictions inherent in our current way of living are resolved then, of course, it is next to impossible to imagine how we might get there. But life isn't like that. When Adam Smith published *The Wealth of Nations* in 1776, no one predicted the rise of the multinational corporation or the shopping mall. Over the next two centuries people explored his ideas in terms of their own lives as they were at the time. Some were in a better position than others to act on their changed perceptions – which may, or may not, have corresponded with Adam Smith's own hopes for the future – but they left few untouched. Ideas have to be tested in the real world. If they do not have the power to move people to change even the smallest action they will hardly galvanise a society to fundamentally rethink where it is going.

When we are in the wilderness it is important to believe in the promised land, but to keep our eyes forever on the far horizon in the hope of spying the land of milk and honey is to court despair and the temptation to stay put. It is far better to put our energies into resolving the difficulties that are in front of us, trying always to ensure that each step we take leads us nearer to our ultimate goal. With that in mind, I thought it might be helpful to complete this little book with some thoughts about what steps it is possible to take today and tomorrow, in the hope that they will help us all to begin the journey.

I have stressed that we are the architects and builders of our own lives but that we are nevertheless constrained by the materials that are to hand, many of which pre-exist us in the form of society's norms and expectations. Although institutions dominate our lives, there are plenty of examples, past and present, in our culture and elsewhere, of ways of living and acting that run counter to this trend and offer focal points around which new life might form and evolve. Today's world is fundamentally sterile but the life force, once it has a toehold, is capable of carrying all before it, splitting the very rocks and bringing forth abundance.

There are three main areas that we should concentrate on and begin to experience in our lives if that change is to happen: first, relationships and personal development; second, token economies and learning to be ourselves; and, third, community building and discovering a sense of belonging. These categories obviously overlap but can be approached quite separately. There are many starting points and, in dealing with each issue in turn, I can only hope to touch the surface. Much of what we already do and think of as being important will come under these headings. Recognising that may help us to find the next step most appropriate to us. The names and addresses of useful organisations are listed at the end of the chapter.

Relationships and personal development

A central theme of this book has been the importance of relationships and, given the fact that many of the problems we encounter in daily life can be traced back to relationships that aren't working, it seems to me that we all have an interest in improving the quality of how we relate one to another. I would go further and say that finding a language, and the corresponding inter-personal techniques to explore ourselves and how we relate to one another, will determine the future shape of humankind. In past ages people could escape their own inner turmoil – not to mention their neighbours – by seeking and colonising new worlds, by fighting battles and wars or by amassing personal wealth and fortune. All such activities focused on the external world. Today, the room for such diversion is limited and becoming more so by the day. The image of 'Earth rising', the finite orb of the Earth silhouetted against the blackness of space, has changed our view of ourselves for ever. There really is nowhere to hide. We finally have to face ourselves and those we relate to. Today's 'New World' is inner space.

Much of my thinking about relationships has come from working with people with learning disabilities and the difficulties of finding individuals to support them who can relate to them as fellow human beings and not as 'things'. I have come to realise that we are all abusive by nature, if only because we are insensitive to other people, even (perhaps especially!) our nearest and dearest. We continually do and say things that hurt others, put them down or effectively deny their existence as autonomous beings – and, for the most part, we are not even aware that we are doing it.

Part of this 'blindness' is undoubtedly due to the complexity and pressure of today's life and the fact that the institutional world inherently denies individuality, preferring to treat people as numbers or categories. As long as we are absorbed in that world we will act according to its laws. But it is equally true that, of all our senses, our awareness of others is among the least developed. An awareness of self and how we come across to others develops through interacting with others in an atmosphere of mutual trust and honesty. If we can share with others how we truly feel, we not only have a chance to recognise that anger, depression, self-doubt, etc. are not in themselves 'bad' but we can begin to put them in some kind of perspective and contemplate strategies for responding to them, thus robbing them of their destructive, irrational element. If other people can also feed back to us how we are perceived – when we appear as negative, judgemental, etc. – we have the opportunity to recognise these traits in ourselves and to do something about them. In short, we can develop as human beings.

We are entering the world of personal responsibility where we begin to dimly perceive that we are inextricably bound up one with another and that, just as we have the right to be treated and accepted as ourselves, so we must extend that privilege to others. That means becoming ever more sensitive both to the needs of others and to our part in any reciprocal transaction. Being able to negotiate on the basis of mutuality means being able to put oneself in the position of the other. A different focus will be required, a 'way of being' that is both more empathic to the other and more firmly grounded in our own true self. We will need to discover more open and honest forms of communication (in which language will play an important, though by no means the most important, part) through the development of active listening, of assertiveness and of reflective skills that mediate between the realities of self and other. We must be prepared to use our whole being in that quest.

Interpersonal skills are something that can be improved and developed: there are many groups and classes available in a variety of settings. Changing how we relate to ourselves and other people requires practice and, ideally, feedback from people we trust and respect. That is a theme I shall return to later. Often, the way we react to the world may have as much to do with our own personal culture as to the wider society that gives expression to our lives. We are, after all, raised in the same general milieu and yet our reaction to the same objective situation can be very different. At a simple level, our attitude to risk will be conditioned by our own experiences and the response of our parents as we were growing up. A family that is very conscious of risk and is always warning a young child against dangers is more likely to produce an adult who sees the perils inherent in any situation. Equally, our ability to relate to others on the basis of mutuality may be limited by our life experiences and the expectations they have created.

We can learn much about the forces that have shaped the way we respond to the world – ways that may not reflect who we truly are as people and which may, therefore, be negative and ultimately life-threatening – with the support of our friends and families. Beyond that there are a range of interpersonal activities in every locality that are more or less explicitly committed to self-development. From psychodrama workshops, self-help groups dealing with a range of specific issues (phobias, alcohol, violence in marriage, etc., etc.), classes in assertiveness, to Tai Chi, massage and a variety of relaxation techniques, the emphasis is very much on broadening the base of one's personal experiences and thereby promoting self-awareness.

To understand and come to terms (without necessarily being able to change) with some of our more deep-seated ways of responding to situations and relationships may require some form of specific counselling. It is not appropriate in this context to discuss the differences between counselling, psychotherapy and psychoanalysis. Suffice it to say that there are many different schools and approaches and some assistance may be required to determine which is the most appropriate in any particular case. Counselling tends to be the most readily available and is provided in a range of settings such as GP surgeries, voluntary organisations and schools as well as in private practice.

The essence of all forms of individual therapy is the relationship, and research shows that, when that is working well, orientation is less critical. There should be some, preferably explicit discussion of the contract you are entering into with a therapist, which may include agreement about the nature of the problem areas to be tackled, the duration, the cost, the boundaries (length of sessions, confidentiality), etc. It is a mutual, if somewhat asymmetric, relationship in which the practitioner should be assisting you to formulate what is essentially your own personal agenda of work in progress, facilitating rather than imposing it. Counselling can stir up strong emotions, emotions we weren't aware we harboured. It can also provide an experience, perhaps the only experience that some people may have had, of the non-judgemental acceptance (which is one expression of love) that is at the base of living 'in community'.

Counselling is still a relatively new area of human endeavour and the standards of training available can be variable. Unless you make contact through personal recommendation it is important to choose a counsellor who is accredited (or working towards accreditation) by the British Association of Counselling: such accreditation will ensure that the counsellor has undertaken at least two to three years of personal development work themselves, vital if they are trying to encourage growth in others!

A greater awareness of and sensitivity to relationships is a prerequisite for creating a saner and more human world. For that future to exist in practice will require change in all of us. Perhaps most significantly it will need a complete reorientation in the way we bring children up and the education we offer them, from an external ordering of more and more measurement and information to the internal experience of ever deeper feeling and understanding. We all have a stake and a part to play in that revolution.

If the effort appears immense, the potential rewards are prodigious. The outpouring of humanity that will result is summed up for me in a passage from St. Matthew's Gospel 25:31–45 when Jesus talks of separating the sheep and the goats.

Then the king will say to those on his right: 'Come, you who have won my father's blessing! Take your inheritance – the kingdom reserved for you since the foundation of the world! For I was hungry and you gave me food, I was thirsty and you gave me a drink, I was lonely and you made me welcome, I was naked and you clothed me. I was ill and you came and looked after me. I was in prison and you came to see me there". Then the true men will answer him, 'Lord, when did we see *you* hungry and give you food? When did we see *you* thirsty and give you something to drink? When did we see *you* lonely and make you welcome, or see *you* naked and clothe you, or see *you* ill or in prison and go to see you?' And the king will reply, 'I assure you that whatever you did for the humblest of my brothers you did for me.'

Rather than being ignorant of how we hurt and destroy one another, we should be seeking to create a world in which we are as unconscious of the help and succour we bestow because it has become second nature.

Token economies and learning to be ourselves

Token economies have existed since the dawn of time. From cowrie shells to the nineteenth-century tokens redeemable in the mill-owner's shop we have used such symbolism as a proxy to expand the limitations inherent in face to face barter. In that sense LETSystems are merely another example of how, in time, all things come full circle. However, just as a wheel has to turn to move the vehicle forwards, so LETSystems offer a fresh starting point and the opportunity of doing some things differently.

There are over three-hundred LETSystems in Britain alone, each varying in its emphasis and interests. I belong to Manchester LETS. It is one of the larger systems, with over five hundred members, offering a wide range of goods and services; from domestic help to computer programming, from aromatherapy to furniture restoring, from tennis to tarot reading. Many businesses have joined, allowing members to pay Bobbins (the local LETS currency) to purchase anything from food in a local café to building work, from legal advice to having your bicycle professionally repaired (it is easy to see how small businesses might actually be stimulated by a LETSystem: baking bread for a regular group of customers could be extended, if the individual wished, into a more full-time commitment). LETShire allows a wide range of members' tools to be loaned out for periods of time. The only limitation is people's imagination and the energy to act on new ideas.

There are two major psychological hurdles that many people find it hard to leap on joining LETS. The first, and easiest to deal with, is the concept of being 'in commitment'. Many people have been brought up to believe that being in debt is 'bad', almost a sin. As a consequence they prefer to sit and wait for someone to pay them for a product or service before looking for something they might want. The result is that nothing happens and they are left feeling let down and discouraged. It has been emphasised earlier, but will bear repetition, that being 'in commitment' is a virtue in LETS because it stimulates trading. And once you have grasped the simplicity of the idea you cease to check whether you're in commitment or credit. You become much more spontaneous in what you purchase and, equally, more flexible in what you offer for trade. You often end up doing a reciprocal trade with someone who has approached you, creating a greater sense of mutuality and increasing the chances of further interactions.

The other reason why people don't trade is the psychological hurdle of being yourself in the transaction. When we go into a shop or phone for a service we don't take account of the other person. And that is true in reverse. Even if we are a regular customer, the shop assistant or tradesperson is unlikely to be much interested in us as an individual. What is on offer is usually a take-it-or-leave-it affair with little room for negotiation. In LETS the emphasis is on a trade between two individuals who both have to explore what they really want and what they are really prepared to offer. There is potentially much more at stake. We are having to be more ourselves.

Trading is also more likely to take place on our home territory rather than in a shop or office and, again, that can mean us having to reveal more of ourselves than we might be used to. We may be happy to display our skills to our nearest and dearest but offering them to comparative strangers is another thing. We might not even recognise that we have skills that people outside our immediate family might want. 'But I've got nothing to trade!' is a frequent refrain with new members. Once again it is an issue of personal development, of taking that first step towards valuing ourselves. It is a good idea to do your first few trades with people that you do know well. When your confidence has grown you will begin to seek out people you may not know. Your network of mutually supportive relationships will begin to grow and, in the process, realising that you do have capabilities that are recognised by others will do wonders for your self-esteem. Once again, a virtuous circle is being created.

One of the major disadvantages of Manchester LETS from my point of view is the geographical area covered. It is very difficult to maintain contact with people on the other side of a great metropolis. The emphasis has to be more local, creating networks of relationships that are self-sustaining and don't just depend on trading. As Manchester LETS continues to grow, the intention certainly is to break down into smaller, more localised patches while retaining the advantages of being able to trade across the whole network for more specialised activities.

Because of my particular interest in developing communities, I have anticipated this development by becoming involved in starting a system in our local Church of England church, involving at most fifty people. Given the history of the early Church it is perhaps surprising how few mainstream churches today are what might truly be called communities. They tend to be Sunday-only congregations with little or no contact in between. Our LETSystem has been a way of breaking some of those barriers down, giving people the opportunity to interact in a different way and throughout the week. For example, by spending half an hour with one of the older members of our congregation, helping her with her windows and sharing a cup of tea, I experienced more genuine communication and deeper understanding than had occurred over years of sharing communion together.

As well as trading goods and services, each member tithes Clems (our church is called St Clement's!) to the church and this allows a variety of work to be paid for in local currency. As a result the building is cleaner and more welcoming. People now look for tasks that need doing rather than wait until they are asked. Many of these activities are done together, perhaps starting with a shared meal, which, once again, provides opportunities for coming alongside one another. We hope that the increasing sense of identity, coupled with the valuing and respect, both of self and others, that comes with having the opportunity to know one another more deeply, will engender a genuine aura of community that will make the church a place where people want to be. Something is happening where before it didn't. That is a tangible example of the small steps that each of us can take and how, together, we can change the world.

The potential of LETS is catching on. I know of a housing association in Manchester that is actively considering allowing tenants to pay at least some of their rent in local currency. The sums collected would then be used to pay local people to carry out repairs and generally keep the estate clean and attractive. Several local authorities are also looking at how the concept might be used to enhance the quality of life in their areas and, in at least one town, discussions are taking place with farmers in the surrounding countryside to see if their produce could be distributed locally through the LETSystem rather than travelling up and down the country in search of a market.

There are many other examples of community-building ventures that can exist either independently of LETS or in conjunction with it; credit unions, bartering organisations, wholefood co-operatives, housing co-operatives, etc., etc. All are experiments in doing things differently. That is what we need. There is no blueprint for an alternative way of living. We have to get on with it, making mistakes and accepting disappointments along the way. The greater the variety the more likely it is that someone or some group will stumble upon and open up a route that the rest of us can follow.

Community building and discovering a sense of belonging

Most of us have had – often intense – experiences of belonging. Whether it be at work, in the family, participating in a sport or other voluntary activity, or just times in our lives when we have felt very close to a group of people, it is a potentially universal sensation. Community building is different only in degree, implying that there is a *conscious* striving to create and sustain that state and there is usually some goal towards which it is directed. Almost any situation where people come together would benefit from community-building and it is therefore surprising that so little attention is paid to it.

One of the reasons for this oversight is that we live in a hierarchical society where responsibility typically resides outside the group. Being 'in community' implies a consensual approach to decision-making with accountability remaining in the group. The members don't necessarily have to be equal in organisational status but they do have to be equal in terms of their participation in the community. Such autonomy can appear threatening to the great and the good.

Having said that, there is growing interest in giving responsibility to work teams. From car production to the army there is a trend towards passing down targets and encouraging the group themselves to work out the best way to achieve them. Teams are potentially mini-communities, and it is by participating in teams that we can develop many of the skills and sensitivities that will be needed in the sufficient communities that have been described in this book.

Anyone who has worked in a team will know that when it is working well anything is possible. You are riding the crest. There is energy to spare, no problem is too great and you look forward to start of each new day. Conversely, when a team is struggling, life can seem like a nightmare, with tiredness and depression endemic and a tendency to want to stay in bed rather than face another day of struggle and discord. The difference between the two states can be due to many factors but a significant cause will always be the participants themselves. It is something about the dynamics in the group and that can be altered if there is the motivation. And what greater motivation can there be for change than hating what you are doing?

Once again, it is a matter of personal development. Only this time the opportunities are built into the system. The more open and honest we can be with our colleagues and the more we can listen to and appreciate their point of view, the more likely it is that we will maximise team functioning *and* achieve a better sense of who we are. In other words, if we can get through the 'chaos' that results from differences in our individual and long-cherished expectations, prejudices and ideologies as well as our need to heal and control, we have the potential for experiencing community.

That can be hard work. Scott Peck has evolved techniques for leading groups into community, and people who have experienced his workshops have found them inspiring. For most situations, however, the skills are not yet sufficiently refined or readily available to make a significant impact. Yet we can take responsibility for our own team/community building, and it doesn't have to involve the dangling off a rock face so beloved of instructors hired by major corporations to teach their top executives how to trust one another!

A first requirement is a genuine desire to improve how a team or community is functioning, and that implies being prepared to change oneself. It is no good starting off with the assumption that the problem lies in someone else and that, if only they would change, everything will be wonderful. The reason you perceive that other person to be the problem may be because you are the problem to them. So there has to be a certain openness to looking at oneself as well as the rest of the group. Once everyone is at least prepared to sign up to that possibility a simple and relatively unthreatening audit of how people feel about the situation can be undertaken. The following are some suggestions about issues that might be explored: they are by no means exhaustive:

1 Identity: who is in this team, does everyone share that perception?
2 Style: where is the team coming from, is it democratic, to whom is it answerable, where does leadership lie, where does power lie?
3 Framework: what is the purpose, what responsibilities do we have, how – and to whom – are we accountable, what are our expectations?
4 Communication: how far do the following words describe the team-negotiation, compromise, flexibility, sharing, participation, ability to disagree, acceptance of criticism, ability to solve problems, listening, summarising, giving feedback, openness? Who in the group tends to contribute each of these skills?
5 Attitudes: are members honest with each other, appreciative of one another's contribution, do they trust each other, value one another, are they reliable, committed, is there humour in the team, optimism, understanding?
6 Support systems: whom can we approach for help and advice, who can give us an outsider's view of how we are doing?
7 Time dimension: how long are we going to function as a team, what opportunities are there for us to get to know one another?

Using such an audit will provide a context and opportunity to discuss personal experiences and, more importantly, the differences between them. The more we question and share, the more experienced we will become at looking at situations and determining what needs to be done to change things for the better. The object is to engage everyone and to be able to utilise their skills, personality and perceptions for the good of the agreed enterprise. It is an essentially consensual response to life's problems, believing that several heads working genuinely together are more likely to come up with an 'outcome of the moment' than any single individual however talented they may be. It is an organic rather than a mechanistic response to life.

Final, final thoughts

The three areas outlined above overlap and complement each other. As we become more practised in the language and techniques they imply we will find them coming ever closer together until it becomes impossible to conceive of advances in one without corresponding changes in the others. Our lives will become less compartmentalised and we will begin to judge all our actions, whether at home, at work or at play, from the same life-enhancing perspective, the perspective that places our humanity at the centre of everything we do. In the process our moral judgement will become more finely tuned and we will begin to feel more in control of our destinies. We will begin to sense where we are going. As we begin to find ourselves part of a group of like-minded souls we will experience ourselves in a different, more whole way. We will have come home, the place where we can truly be ourselves. And each group, by its example, will offer inspiration to others. The individual seeds will flourish and bring forth a great crop. Which, in turn, will provide an even greater store of seed for the future.

If there is a message in this section of the book it is above all 'do'. Look for opportunities in your life, your workplace, your neighbourhood to reclaim the human. Don't be afraid to try new things and don't be afraid of having to change yourself.

Remember, each small step that any of us takes is potentially a giant leap forward for humankind.

NOTES

Chapter 1 *The Pursuit of Happiness*

1 A. H. Maslow, *Motivation and Personality* (Harper & Row, London, 1987).

The five elements in Maslow's Hierarchy are;
1) Physiological needs: hunger, thirst, reproduction, etc.
2) Need for safety: security, stability, order, etc.
3) Need for love: belongingness, affection, affiliation, identification, etc.
4) Need for esteem: prestige, success, self-respect, etc.
5) Need for self-actualisation: a realisation of one's full potential

2 Carl Rogers, *On Becoming a Person* (Constable, London, 1988), p. 203.

3 Colin Turnbull, *The Mountain People* (Pan Books, London, 1974).

4 P. Kropotkin, *Mutual Aid* (Penguin Books, Harmondsworth, 1939), p. 217

5 The schema used in this section derives from the work of N. Elias (author of *Über den Prozess der Zivilisation*, Basle, 1939) and, specifically, from E. Dunning, 'Two Illustrative Case Studies', in P. I. Rose, *The Study of Society* (Random House, New York, 1967).

6 L. Mumford, *The Myth of the Machine* (Secker & Warburg, London, 1967) and L. Mumford, *The Pentagon of Power* (Harcourt Brace Jovanovich, New York, 1970).

7 G. Orwell, *Nineteen Eighty-Four* (Penguin, Harmondsworth, 1955), p. 215.

8 C. G. Jung, *Man and His Symbols* (Pan Books, London, 1983), p. 71.

Chapter 2 *The Unholy Alliance*

1 M. V. C. Jeffreys, *Personal Values in the Modern World* (Penguin Books, Harmondsworth, 1966), p. 149.

2 George Bush's 1988 Presidential Election pledge.

3 Erich Fromm, *To Have or To Be* (Abacus, London, 1990), p. 141.

4 *Psychology and Alchemy*, para. 8, as quoted in F. Fordham, *An Introduction to Jung's Psychology* (Penguin Books, Harmondsworth, 1972), p. 74.

5 Lapace's remark to Napoleon.

6 The scientific method was sketched out in its essentials by Descartes in the first half of the seventeenth century. His first principle effectively separates the universe into subject (me – in here) and object (the world - out there)

as being the only basis for the subject – I – to study the object - the world, including you - from a detached, non-involved point of view. The principle of objectivity requires us to accept that the world is entirely out there, whereas common sense tells us that the world is only out there in the sense that we, ourselves, are able to observe it. If humans didn't exist, the world might still be there but it wouldn't be known in the same way. In other words, what we know about the world exists *within* us. Descartes' objectivity is a pose – a useful one in certain circumstances, but a pose none the less. The problem is that we now take it as the real thing. Detachment at all costs! It is a view that, taken to extremes, is as morally bankrupt as the deal-of-the-moment morality of the money economy.

7 E. F. Schumacher, *A Guide for the Perplexed* (Abacus, London, 1984), p. 45.

Chapter 3 *Understanding the Question*

1 E. F. Schumacher, *A Guide for the Perplexed* (Abacus, London, 1984), p. 126.
2 A. Koestler, *The Ghost in the Machine* (Picador, London, 1967), p. 48.
3 David Wade, *Crystal & Dragon* (Green Books, Devon, 1991), p. 75.
4 R. Higgins, *The Seventh Enemy* (Hodder & Stoughton, London, 1982), p. 254.
5 T. S. Eliot, *On Poetry & Poets* (Faber & Faber, London, 1957), p. 7.
6 R. E. Hobson, *Forms of Feeling – The Heart of Psychotherapy* (Routledge, London, 1994)
For example: 'The skill of a psychotherapist lies in his ability to learn the language of his patient, and to help in creating a mutual language – a personal conversation' (p. 46). A little later he emphasises,: 'Language grows. It is continually transformed. The multiplicity of forms of life is not some-thing that is fixed, given once and for all. New language games emerge and others become obsolete, die, and are forgotten. The process of death and rebirth of language is the story of development of the individual and of culture'(p. 48). In other words, language is the medium through which change is initiated, embraced and come to terms with.
7 E. F. Schumacher, *Small is Beautiful* (Abacus, London, 1983), p. 98.
8 E. F. Schumacher, *A Guide for the Perplexed* (Abacus, London, 1984).
9 *Ibid.*, p. 142.
10 K. Popper, *The Open Society and its Enemies* (Routledge, London, 1966).
11 'A Friendly Approach to Change', *The Observer*, 25 April 1995.
12 M. Scott Peck, *A World Waiting To Be Born* (Arrow, London, 1994).
13 *Ibid.*, p. 347.

Chapter 4 *Freedom, Equality and Love*

1 E. F. Schumacher, *Small is Beautiful* (Abacus, London, 1983), p. 129.
2 I Corinthians 13:4–7.
3 Luke 19:11–27.

4 *New Internationalist* (September 1985), p. 5.
5 *New Internationalist* (May 1987), p. 25.

Chapter 5 *Building a Sufficient Community*
1 Erich Fromm, *The Sane Society* (Holt, Rinehart & Winston, New York, 1955).
2 M. Scott Peck, *The Different Drum* (Arrow, London, 1990), p. 126.
3 'Taming the Dragon', *Brass Tacks*, BBC2, 8 October 1987.
4 Which then identify totally with the product. After the Gulf War, inhabitants of St Louis, Missouri (home of McDonnell Douglas) were ecstatic about how their equipment had performed and could describe film sequences of Cruise missiles exploding in graphic detail. 'That's my job, to produce the best weapons system there is.'
5 Thornton Wilder.
6 M. Scott Peck, *The Different Drum* (Arrow, London, 1990), p. 86.
7 *Ibid.*, p. 94.
8 *Ibid.*, p. 102.
9 E. F. Schumacher, *Small is Beautiful* (Abacus, London, 1983), p. 98.
10 M. Scott Peck, *A World Waiting To Be Born* (Arrow, London, 1994), p. 363.

Chapter 6 *The Sufficient Community in Practice*
1 J. Porritt, *Where on Earth are We Going* (BBC Books, London, 1990), p. 41.
2 M. Andrews, *The Birth of Europe* (BBC Books, London, 1991), p. 179.
3 J. Lovelock, *The Ages of Gaia* (OUP, Oxford, 1990), p. 232.
4 P. Nuttgens, *The Home Front* (BBC Books, London, 1989), p. 67.
5 *Ibid.*, p. 69.
6 *Ibid.*, p. 10.
7 J. Lovelock, *The Ages of Gaia* (OUP, Oxford, 1990), p. 232.
8 P. Nuttgens, *The Home Front* (BBC Books, London, 1989), p. 41.
9 E. F. Schumacher, *Small is Beautiful* (Abacus, London, 1983), p. 98.
10 *Ibid.*, p. 124.
11 M. Proust, *The Germantes Way*, quoted in introduction to M. Proust, *Pleasures and Regrets* (Grafton Books, London, 1988), p. x.

USEFUL ADDRESSES

The British Association of Counselling, 1 Regent Place, Rugby CV21 2PJ
(Tel. 01788 578328)
(operates a code of ethics and practice for accredited counsellors)

LETSlink, 61 Woodcock Road, Warminster, Wiltshire BA12 9DH (Tel.
01985 217871)
(co-ordinating information about LETSystems around the country)

LETSolutions, 124 Northmoor Road, Longsight, Manchester M12 5RS
(Tel. 0161 434 8712; Fax. 0161 257 3686; e-mail:
StevenKnight@compuserve.com)
(offering consultancy and advice about the setting up of LETSystems)

Intermediate Technology Development Group, Myson House, Railway
Terrace, Rugby CV21 3HT (Tel.: 01788 560631)
(advice and examples of forms of appropriate, people-centred technology)

The Centre for Alternative Technology, Machynlleth, Powys SY20 9AZ
(Tel. 01654 702400)
*(interesting centre to visit, provides advice and can custom-build equipment to suit
individual needs)*

The Foundation for Community Encouragement, P.O. Box 449, Ridge-
field, Connecticut 06877, USA (Tel. 203 431 9484)
(promotion of Scott Peck's work)

Resurgence, Subscription Department, Rocksea Farmhouse, St. Mabyn,
Bodmin, Cornwall PL30 3BR (Tel. 01208 841824)
(bi-monthly magazine devoted to ecological and spiritual thinking)

INDEX